T0252817

Routledge Re........

Themes in Geographic Thought

Themes in Geographic Thought, first published in 1981, explores in breadth and depth the interrelationships among the history of Geography, geographic thought, and methodology, specifically focusing on the interactions between geographical research and various contemporary philosophical schools: positivism, pragmatism, functionalism, phenomenology, existentialism, idealism, realism and Marxism.

An attempt is made to synthesise Geography's historically rich tradition with the current diversity in approaches to the discipline, based on the belief that 'geographic thought', at any point in time, is a manifestation of the mutual influence between the prevailing philosophical viewpoints and the major methodological approaches in vogue. Each chapter presents an overview of the concrete ideas of a particular school of philosophy and stresses its relevance and impact on various aspects of Geography.

Themes in Geographic Thought

Edited by
Milton E. Harvey and Brian P. Holly

Routledge
Taylor & Francis Group

First published in 1981
by Croom Helm Ltd

This edition first published in 2014 by Routledge
2 Park Square, Milton Park, Abingdon, Oxon, OX14 4RN
and by Routledge
711 Third Avenue, New York, NY 10017

Routledge is an imprint of the Taylor & Francis Group, an informa business

A Library of Congress record exists under LC control number: 85673625

ISBN 13: 978-0-415-74748-6 (hbk)
ISBN 13: 978-1-315-79703-8 (ebk)
ISBN 13: 978-0-415-74750-9 (pbk)

Themes in Geographic Thought

Edited by
MILTON E. HARVEY and BRIAN P. HOLLY

CROOM HELM   LONDON

© 1981 M.E. Harvey and B.P. Holly
Croom Helm Ltd, Provident House, Burrell Row,
Beckenham, Kent BR3 1AT
Reprinted 1983

British Library Cataloguing in Publication Data

Themes in geographic thought. – (Croom Helm series in geography
and environment).
1. Geography – Methodology
I. Harvey, Milton E.   II. Holly, Brian P.
910'.01   G70

ISBN 0-7099-0188-7

Printed and bound in Great Britain by
Professional Books, Abingdon, Oxon

TO LYNDA AND MARY LOU

CONTENTS

PREFACE

The idea for this book arose out of our mutual interest in the interrelationships among the history of Geography, geographic thought and methodology. The two of us have been responsible for teaching undergraduate and graduate level courses in research methods and geographic thought for almost ten years. This experience has underscored the need for a comprehensive book treating the fundamental philosophical themes in our discipline. It became apparent to us that no college level text existed which attempted to synthesise the relationship between Geography's rich historical tradition and the current diversity in approaches to geographic research. Too often 'geographic thought' has been equated with the history of Geography and geographical ideas. We feel that this approach does not adequately represent thought in Geography.

While this volume makes no claim at being comprehensive, it does attempt to present the major philosophical positions which have influenced geographical research, at least in the West. We felt that so eclectic an undertaking such as this book was beyond the abilities of a single or even two authors. Therefore, we settled on an approach whereby major philosophical 'schools' would be presented by persons closely identified with them in the field. We believe that the spatial theme will persist in Geography and that the major themes in geographic thought will be philosophical.

As originally conceived, this book was to be used as an advanced undergraduate and graduate level text in Geographic Thought courses and as a basic reference work on contemporary philosophical themes. Our only charge to the authors of the individual chapters was that they present an overview of the concrete ideas of the philosophy, stress its relevance and impact on various aspects of Geography and present the philosophy's relationship to other current philosophical positions in Geography.

Several persons deserve mention in connection with the effort involved in producing this book. For his patience with our delays and his insistence that this book be completed, we extend our apologies and thanks to Peter Sowden of Croom Helm. Surinder Bhardwaj, Stavros Constantinou, Susan Despres and Mary Ann Ulan also deserve thanks for reading and commenting on various parts of the manuscript.

Preface

Our appreciation is also extended to Della Carpenter for the application of her considerable typing and editorial skills.

M.E.H.
B.P.H.

1 PARADIGM, PHILOSOPHY AND GEOGRAPHIC THOUGHT

Milton E. Harvey and Brian P. Holly

Introduction

Geographic thought, at any point in time, is a manifestation of the interaction between the prevailing philosophical viewpoints and the major methodological approaches in vogue. Because of the extreme diversity of viewpoints on both philosophy and methodology, there has been a constant extension, and even a shift, in the focus of the discipline. In the last few decades many papers, monographs and books have been written either about specific aspects of these shifts or about the major trends in these shifts. Ley and Samuels' *Humanistic Geography* (1978) and Peet's *Radical Geography* (1979), are examples of the former. Whereas Hartshorne's *Nature of Geography* (1961), James's *All Possible Worlds* (1972) and James and Martin's *The AAG: The First Seventy-Five Years* (1979) are examples of the latter. In different forms, all these authors have attempted to organise and present specific viewpoints on the history, philosophy, the focus and the methodology of geography. Pervading all these are attempts to clarify the influence of philosophical and paradigmatic viewpoints on geographic thought. Because of the pedagogical aim of this book, the first part of this chapter will explore this interrelationship. Later, we discuss the major paradigmatic developments in geography. In the final section, the trends toward strong philosophical viewpoints within emerging geographic paradigms are explored.

Paradigm, Philosophy and Thought

Ever since David Harvey's paper on revolution and counter-revolution in 1972, the use of the word paradigm has become fashionable in geography as well as having become a pivotal concept for courses in geographic thought on both sides of the Atlantic. Thomas Kuhn has become as familiar to students of geography as Hartshorne or Humboldt! Kuhn, in his classic book, *The Structure of Scientific Revolutions*, defined paradigm as 'the entire constellation of beliefs, values, techniques, and so on shared by the members of a given community' (1970a, p.175). In an earlier articulation of paradigm (the 1962 edition of *The Structure of Scientific Revolutions*), Kuhn uses this concept in at least 21 different ways (see Masterman, 1970, pp.61-5). These were collapsed into three paradigm types by Masterman: the metaphysical paradigms or metaparadigms, the sociological paradigms and the artefact or construct paradigms. Very briefly, the meta-paradigms present a total global view of science. It is a gestalt view; a *Weltanschauung* (or map). Such a map, Ritzer wrote, 'allows the scientist to explore the vast and complex world otherwise inpenetrable were he to explore randomly' (Ritzer, 1975, p.4). Such a paradigm performs three basic functions:

1. It defines what entities are (and are not) the concern of a particular scientific community.
2. It tells the scientist where to look (and where not to look) in order to find the entities of concern to him.
3. It tells the scientist what he can expect to discover when he finds and examines the entities of concern to his field (Ritzer, 1975, p.5).

It is the broadest area of consensus in a discipline, and it defines the subareas of research. In contrast, the sociological paradigm is grounded on concrete scientific achievement; a universally recognised scientific achievement. The third, and narrowest, is the construct paradigm. Here specific entities such as a textbook, an instrument, or a classic work are viewed as paradigms. Conceptually, the construct paradigm is largely subsumed in the sociological paradigm and this, in turn, is essentially subsumed in the metaparadigms. Because of such a nesting, the construct paradigm must be central to the development of paradigmatic sciences. In fact, Masterman argues that it is the construct paradigm that is central to Kuhn's formulation:

> If we put what a Kuhnian paradigm is, Kuhn's habit of multiple definition poses a problem. If we ask, however, what a paradigm does, it becomes clear at once . . . that the construct sense of 'paradigm,' and not the metaphysical sense of metaparadigm, is the fundamental one. For only with an artefact can you solve puzzles . . . It remains true that for any puzzle which is really a puzzle to be solved by using a paradigm, this paradigm must be a construct, an artefact, a system, a tool; together with the manual of instructions for using it successfully and a method of

interpretation of what it does (Masterman, 1970, p.70).

In his 'Reflection on My Critics', Kuhn agrees with Masterman that construct paradigms are central to his thesis. 'If I could', he noted, 'I would call these problem-solution paradigms, for they are what led me to the choice of the term in the first place' (Kuhn, 1970c, p.277). Because of the subject-specific nature of problem-solutions in the framework of the construct paradigm, Kuhn substituted the phrase 'disciplinary matrix' for 'a paradigm' or 'a set of paradigms'. He amplified his preference for disciplinary matrix thus:

> 'disciplinary', because it is common to the practitioners of a specified discipline; 'matrix,' because it consists of ordered elements which require individual specification. All of the objects of commitment described in my book as paradigms, parts of paradigms or paradigmatic would find a place in the disciplinary matrix, but they would not be lumped together as paradigms, individually or collectively (Kuhn, 1970c, p.271).

This discussion of the concept of paradigm creates the proper backdrop for a discussion of paradigms in geography. However before such an exercise can be fruitfully attempted, a few other issues pertinent to this concept must be explored. These are posed as questions: Is a scientific discipline, by definition, a single paradigm? How are new paradigms created? Is a paradigm equivalent to a theory? Is a paradigm associated with a specific philosophy?

Single or Multiple Paradigms

On the question of single or multiple paradigm disciplines, Kuhn's earlier formulation was unclear. Masterman wrote thus: 'As I see it, he fails to distinguish from one another these relevant states of affairs, which I will call respectively non-paradigm science, multiple-paradigm and dual-paradigm science' (Masterman, 1970, p.73). Briefly, a discipline is in a non-paradigmatic state when there are no paradigms and the scientists in a particular discipline cannot differentiate the subject matter of their discipline from that of other allied disciplines. In contrast, a dual-paradigmatic science exists just prior to a revolution which leads to the emergence of a single paradigm. During this period, the two paradigms compete for control. In a stages framework, a discipline moves from the non-paradigmatic state through the competitive dual-paradigmatic stage to the single-paradigm stage.

This simple evolutionary framework is, however, disrupted by the possibility that certain disciplines, especially in the social sciences, may be multiple-paradigm sciences. In a multiple-paradigm science, many paradigms are competing for hegemony in the field. As Ritzer observed, 'one of the defining characteristics of a multiple-paradigm science is that supporters of one paradigm are constantly questioning the basic assumptions of those who accept other paradigms' (Ritzer, 1975, p.12). Masterman has vividly summarised the attributes of such a science:

> Here, within the sub-field defined by each paradigmatic technique, technology can sometimes become quite advanced, and normal research puzzle-solving can progress. But each sub-field as defined by its technique is so obviously more trivial and narrow than the field as defined by intuition, and also the various operational definitions given by the techniques are so grossly discordant with one another, that discussion on fundamentals remains, and long-run progress (as opposed to local progress) fails to occur (Masterman, 1970, p.74).

Masterman's discussion of multiple paradigms has been criticised on the grounds that she regards them as a stage in the emergence of a paradigmatic science. Many social scientists contend that paradigms do co-exist in a discipline. Merton recently defended this co-existency on the ground that plurality of paradigms does provide the atmosphere for the generation of,

> a great variety of problems for investigation instead of prematurely confining inquiry to the problematics of a single, assumedly overarching paradigm . . . The exclusive adherence of a scientific community to a single paradigm, whatever it might be, will preempt the attention of scientists in the sense of having them focus on a limited range of problems at the expense of attending to others (Merton, 1976, pp.138-9).

Rather than social scientists, such as sociologists, striving for a unified paradigm, Merton rightly argues that energies should be directed toward the identification of the capabilities and weaknesses of each paradigm. A similar view about the need for this plurality of viewpoints has been stressed by Bartels:

> Plurality of points of view at any one time is . . . unavoidable in a
> discipline and our reaction to this should not be one of dismay
> but a recognition of the need for a corresponding expansion of
> democratic forms of pluralistic co-existence in science which
> accepts these situations of conflict between different 'statements
> of truth' (Bartels, 1973, p.24).

Bartels, however, cautioned that such a plurality should not exist to
such a degree that the subdisciplines are closer to neighbouring
disciplines. Furthermore, he argued that the existence of such
plurality should not prevent the continued dissemination of a clearly
defined public image of that discipline. When both occur, the
discipline is in deep trouble!

From the above discussion, it is evident that there is no consensus
about the stages in the paradigmatic evolution of a scientific discipline.
In fact, it may even be argued that a science may move from a pre-
paradigmatic to a multiple-paradigmatic science, and from the multiple
paradigm state to either a single- or a dual-paradigm science. It is
probably impossible to develop a staged unidirectional formulation for
the evolution of any discipline, particularly the social sciences, and we
do not intend to make such an attempt. Of more relevance to us is
the question of how to determine whether a science is paradigmatic.

Ritzer has identified certain attributes of a paradigm: an exemplar,
image of the subject matter, theories and methods (Ritzer, 1975,
pp.25-7). As defined by Kuhn himself, exemplars are:

> The concrete problem-solutions that students encounter from the
> start of their scientific education, whether in laboratories, on
> examinations, or at the ends of chapters in science texts. To these
> shared examples should, however, be added at least some of the
> technical problem-solutions found in the periodical literature that
> scientists encounter during their post-educational research careers
> and that also show them by example how their job is to be done
> (Kuhn, 1970a, p.187).

To Kuhn, exemplars are those fundamentals learned by the
practitioners of that discipline or viewpoint. The image of the subject
matter is the single overriding theme that is most characteristic of the
dominant exemplar; it is the basic subject matter of that viewpoint.
In addition to exemplar and image of the subject matter, a paradigm
must also include a constellation of theories and methods. An

example from Ritzer's discussion of the social facts paradigm will
help elucidate on these components.

> Exemplars: Durkheim's *The Rules of Sociological Method* (1938)
> and *Suicide* (1951)
> Image of the Subject Matter: Social Structure and Social
> Institutions
> Theories: Structural-Functionalism
> Methods: Questionnaires and/or Interviews

These attributes of a paradigm are later used in our discussion of
paradigms in geography.

How are New Paradigms Created?

In the traditional Kuhnian formulation, it is not the accumulation of
knowledge that causes changes in science. Such changes are caused by
a revolution. In this formulation change is effected through a linkage
of events: Paradigm A → Normal Science → Anomalies → Crisis →
Revolution → Paradigm B. Briefly, after a paradigm has emerged, there
is generally a period of normal science when scientists working within
that paradigm accumulate knowledge. This research expansion results
in the gradual accumulation of anomalies which cannot be explained
or solved by the existing paradigm. As these increase, a crisis stage is
reached as discontent with the paradigm mounts. Ultimately, this
results in a revolution. When comparing his views on scientific
revolutions to those of Sir Karl Popper, Kuhn clearly reiterated this
approach to scientific progress:

> Both of us reject the view that science progresses by accretion;
> both emphasize instead the revolutionary process by which an
> older theory is rejected and replaced by an incompatible new one;
> and both deeply underscore the role played in this process by the
> older theory's occasional failure to meet challenges posed by logic,
> experiment, or observation (Kuhn, 1970b, pp.1-2).

As David Harvey rightly noted, the revolutionary process stressed
by Kuhn and Popper only gains acceptance if 'the nature of the
social relationships embodied in the theory are actualized in the
real world' (Harvey, 1972, p.4).

Paradigm and Theory

In the footnote designed to elucidate on his use of theory in the above quotation, Kuhn wrote thus: 'elsewhere I use the term "paradigm" rather than "theory" to denote what is rejected and replaced during scientific revolutions' (Kuhn, 1970b, p.2). But as Masterman observed, paradigm and theory are not the same! She contended that the meta-paradigm and the sociological paradigm 'are prior to theory' (Masterman, 1970, p.66). Such paradigms precede or are combinations of law, theory and methodology. In contrast, the construct paradigm, Masterman asserts, can be less than or equal to a theory. The distinction between theory and paradigm is also noted by Ritzer: 'theories are not paradigms, but only one aspect of a far broader unit that is a paradigm' (Ritzer, 1975, p.20). In summary, although some theories can be equivalent to paradigms (as in the physical sciences), generally, this is not the case. The broad relationships between paradigms and theory are summarised below:

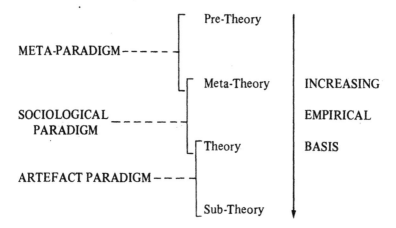

Paradigm and Philosophy

Broadly, an individual's philosophy is the totality of his basic beliefs and convictions. Titus has presented the following definitions of philosophy (Titus, 1964, pp.6-8):

i. It is a personal attitude toward life and the universe. In amplifying this, he wrote that 'philosophy is in part the speculative attitude that does not shrink from facing the difficult and unsolved problems of life' (Titus, 1964, p.6).

ii. It is a method of reflective thinking and logical inquiry. It is the
attempts by the individual to logically understand a specific problem.
It involves a critical evaluation of the facts.
iii. It is an attempt to develop a view about the whole system.
iv. It is the 'logical analysis of language and the clarification of the
meaning of words and concepts' (Titus, 1964, p.8).

A better insight can be gained about philosophy and philosophical
thinking when it is contrasted with science. To quote from Titus:

> Philosophy . . . attempts to gain a more comprehensive view of
> things. Whereas science is more analytic and descriptive in its
> approach, philosophy is more synthetic or synoptic, dealing
> with the properties and qualities of nature and life as a whole.
> Science attempts to analyse the whole into its constituent
> elements or the organism into organs; philosophy attempts to
> combine things in interpretative synthesis and to discover
> the significance of things (Titus, 1964, p.97).

The above discussion underscores the functions of philosophy as
a speculative, descriptive, normative and analytical discipline that
investigates the presuppositions and the scientific work of
practitioners in a discipline. Philosophy evaluates what has been
done, and from there, it may suggest what should be studied by a
discipline. This implies that philosophy not only evaluates but also
creates a framework for research. It is an evaluation of how the
discipline has conducted research; what questions the researchers
have asked, and how these questions have been investigated
within the framework of both normative and societal structures.
In this context, Harris's comment is appropriate:

> To reveal what criteria are presupposed is not by itself
> sufficient. It is further necessary to find out whether they are
> mutually consistent, and, if not, how they require modification.
> When this is determined we say not only what is presupposed
> in fact but what ought to be presupposed to make our thought
> and action consistent (Harris, 1969, p.7).

Central to any philosophical viewpoint is the individual's belief
system. Because of the diversity of beliefs, there is a diversity of
philosophies; whether in the commentaries made on what is being

done or regarding what should be done and how it should be done. This implies that philosophies are not equivalent to paradigms. A set of philosophies with sufficient communality of focus and approach, and a large group of practitioners, may constitute a paradigm. This is especially true of the multifaceted, research-topic dominated social sciences.

In any discipline, philosophies are essential because they create a set of subjectively imposed constraints within which the individual does research. David Harvey emphasised this point when he noted that in conducting research, 'we cannot . . . proceed without some objective, and defining an objective amounts, however temporarily, to assuming a certain philosophical position' (Harvey, 1969, p.5).

Closely associated with both philosophy and paradigms is *methodology*. Stated briefly, a methodology is the logic of an explanation, and a methodologist is concerned primarily 'with the logic of explanation, with ensuring that our arguments are rigorous, that our inferences are reasonable, that our method is internally coherent' (Harvey, 1969, p.6). Thus the methodologist is concerned with the logical nature of an explanation, and the philosopher is concerned with thinking about our 'beliefs with respect to the nature of geography' (Harvey, 1969, p.6). Regarding the level of abstraction or 'detached contemplation', the philosophers are the most abstract, followed by theoreticians, and then the methodologists (Abler, Adams and Gould, 1971, p.4).

In Figure 1.1 we have presented a generalised pattern of the relationships between methodology, philosophy, theory and paradigm. All disciplines are concerned with both explanation and prediction. In the process of explanation, philosophy comes into play in critically assessing the objectives of the explanation and how they are achieved. Theory is used to create the bases for the explanation. For example, in geography, spatial theories such as central place theory and spatial location theory, constitute the bases for the explanations of spatial distributions, spatial interactions and spatial organisation. Methodology of the discipline is the logic used in the explanation. The major philosophical, methodological and theoretical viewpoints in a discipline at any given time constitute the thought of that discipline. Whether that discipline is paradigmatic or not depends on the degree of dominance of any specific viewpoint or viewpoints. In the next section we explore the bases for the existence of paradigms in geography.

Figure 1.1: Thought, Paradigm and Explanation

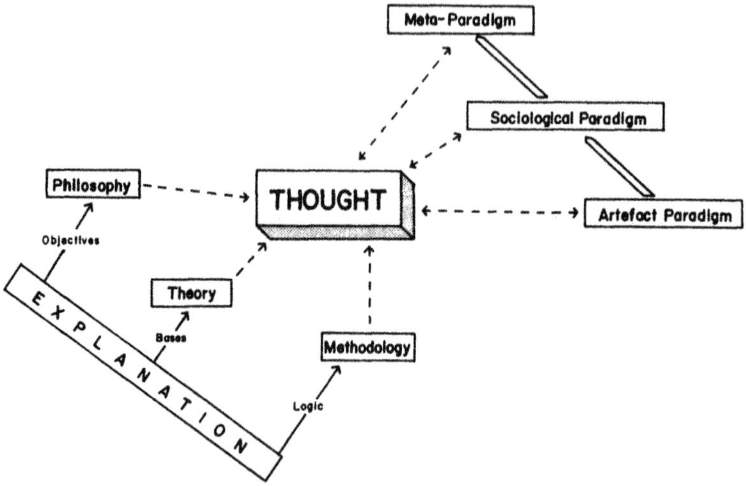

Geography as a Paradigmatic Discipline

In the geographic literature, many scholars have assumed that geography is a paradigmatic science. In fact, some have argued that our discipline has undergone the metamorphosis, the revolution that signals the demise of one paradigm and the creation of another. We need to compare geography as a discipline against the paradigmatic criteria discussed earlier to see if these contentions are valid.

In tracing the evolution of American geography, James and Martin noted four desiderata for the emergence of a field of learning as a learned profession:

> (1) a number of scholars actively working on related problems, and in close enough contact with each other that ideas are quickly disseminated, and critical discussion stimulated; (2) departments in universities offering advanced instruction in the concepts and methods of the field; (3) opportunities for qualified scholars to find paid employment in work related to the profession; and (4) an organization, such as a professional society, to serve the interests of the profession and provide a focus for professional activities (James and Martin, 1979, p.8).

As James and Martin noted elsewhere in their book, such patterns of learned-discipline formation took place in Germany, France and

Britain as well. But learned disciplines, with a specific regional flavour, do not constitute paradigms. They are, of course, a necessary prelude to paradigm formation. As noted earlier, a central focus to the formation of a paradigm is the existence of an exemplar or exemplars. Consequently, we start by identifying what we regard as geographic exemplars. Subsequently, we will critically assess their paradigmatic status.

Geographic Exemplars

In the selection and discussion of what we regard as examplars, we follow very closely the discussion and characteristics of exemplars given by Ritzer and summarised earlier in this chapter. Historically, we will limit our discussion to the late nineteenth and twentieth centuries.

Ratzel's Anthropogeographie. The latter half of the nineteenth century was, in part, dominated by the works of Friedrich Ratzel. Most of his ideas on human geography were published in 1882 under the title of *Anthropogeographie.* Although he modified his ideas in the second edition, the ideas in that 1882 edition influenced the development of geography for at least half a century. With a doctorate in zoology, geology and comparative anatomy, he viewed geography as 'the connection between the natural sciences and the study of man' (Hartshorne, 1961, p.90). He advocated the use of scientific methods in the study of human geography. As Schmidt rightly concluded, one of Ratzel's greatest contributions was 'to have brought this part of the geography of man, cultural geography, into a scientific system by organization of the phenomena, and establishment of concepts and significant connections of the results obtained' (Peter Schmidt, from *Wirtschaftsforschung und Geographie* quoted by Hartshorne, 1961, p.90). In *Anthropogeographie* Ratzel used this deductive approach to present the first systematic study of the geography of man. The first edition of that book was organised 'largely in terms of the natural conditions of the earth, which he studied in their relations to human culture' (Hartshorne, 1961, p.91). This environment-dominated-man-response approach was reversed in his second volume. For details of Ratzel's changing views see Wanklyn (1961), Dickinson (1969) and Sauer (1971). This latter approach, with useful modifications by Hettner and Schlüter, did influence the development of Richthofen's chorological approach, an approach that became associated with twentieth-century German geography. Elsewhere, especially in the United States, Ratzel's earlier viewpoint in the first edition of *Anthropogeographie*, was extended and elaborated by scholars such as Huntington, Semple and Davis into what became known as environmental determinism; a view which dominated American geography for decades. As Grossman recently observed:

> The scientific milieu in the latter half of the nineteenth
> and early twentieth centuries was dominated in part by
> Darwinian ideas, deductive approaches, and an acceptance of
> the concept of Newtonian cause and effect relationships. Fitting
> well into this intellectual environment, the theme of environmental
> determinism, developed mostly by geographers, was the prevailing
> view in American geography at the turn of the twentieth
> century . . . The concern was with documenting the control or
> influence of the environment upon human society (Grossman,
> 1977, p.127).

In America, environmental determinism was widespread indeed. William Morris Davis, for example, stated that geography was concerned with the analyses of the relationships between inorganic control and organic response. As he put it:

> any statement is of geographic quality if it contains a reasonable
> relation between some inorganic element of the earth on which
> we live, acting as control, and some elements of the existence or
> growth or behavior or distribution of the earth's organic
> inhabitants, serving as a response (Davis, 1906, quoted by James,
> 1972, p.359).

Similar views were expressed by others. To Semple, man, like other organisms, was a product of the earth, whereas to Huntington, weather and climate influenced the trends in human history.

Central to all these works on the effects of the inorganic controls on organic response was the notion of Newtonian cause and effect relationships. As James and Martin noted, 'throughout the first twenty years of Association geography . . . (1904-1923), exemplified in the presented papers, was the ubiquitous theme of the causal notion . . . "Influence" gave way to "adjustment", and a variety of ill-defined "determinisms" emerged' (James and Martin, 1979, p.51).

The influence of Ratzel was not just restricted to the initiation of the environmental deterministic approach, he made contributions in the field of political geography that are still central to this subfield. Concepts such as lebensraum, and the organismic structure of the state featured prominently in his *Politische Geographie* published in 1897. He also made contributions to historical geography. Sauer refers to his *Culturgeographie* as 'historical geography of grand design' (Sauer, 1971, p.253).

To summarise, we believe that Ratzel's *Anthropogeographie* was a seminal work and the amount of intellectual debate it created on both sides of the Atlantic makes it an exemplar. Ratzel's views about geography dominated for decades – a tribute to his ability as a teacher and scholar. As Semple wrote,

> He grew with his work, and his work and its problems grew with him. He took a mountaintop view of things, kept his eyes always on the far horizons, and in the splendid sweep of his scientific conceptions sometimes overlooked the details near at hand. Herein lay his greatness and his limitations (Semple, 1911, p.vi).

Vidal's Tableau de la Geographie de la France. Equally central to the development and evolution of geography during the early part of this century was the work of the French geographer Paul Vidal de la Blache. Unlike the inorganic control-organic response basis of the environmental determinists, Vidalian tradition minimised the influences of the environment. Originally a historian, Vidal read the works of Humboldt and Ritter, and later tested his ideas on human geography against those of Ratzel (see Febvre, 1932, p.18). Initially, Vidal's ideas on human geography were presented in a series of articles, 'at once practical and critical, in a rather precise style, with sudden illuminations like flashes of divination and understanding – and with what power of suggestion and even of inspiration all through' (Febvre, 1932, p.18). Central to Vidal's work were the life styles (*genres de vie*) that developed in different geographic environments. As Buttimer noted 'genres de vie, the products and reflections of a civilisation, represented the integrated result of physical, historical, and social influences surrounding men's relation to milieu in particular places' (Buttimer, 1971, p.41).

Vidal's ideas of *genres de vie* were fully developed in his renowned *Tableau de la geographie de la France* (1903), in his *La France de l'est* (Paris, 1917), and in his *Principles de Geographie Humaine* (1922). Specifically, his concept of natural milieu, which evolved from his studies of rural communities, stressed the relationship between the 'inorganically integrated and biotic infrastructure of human life on earth' (Buttimer, 1971, p.45). He linked his *genres de vie* and the natural milieux through the concept of the *milieux de vie* (the adaptation of the natural resources of the milieux by different people). He suggested, on a super-continental scale, that global population distributions be studied through an investigation of how these *milieux de vie* result in the evolution of various and differing *genres de vie*.

Closely related to the above, was his belief that social geography should concentrate on understanding how both the biotic and physical conditions become manifest in the social life of the various societies. Thus, in explaining the differences between groups in the same or similar environments, he relied not on the dictates of the physical environments, but on the attitudes, values and habits of the communities. He argued that changes in attitudes, values and habits did create numerous possibilities for the human communities. His 1902 lecture (quoted by Febvre) implied this contention:

> It must be remembered that the force of habit plays a great part in the social nature of men . . . It often happens that amongst the geographical possibilities of a country there are some obvious ones which have remained sterile or have only been exploited at a late period. We must ask ourselves, in such cases, whether they were in harmony with the manner of life which other qualities or properties of the soil had already caused to take root there (Febvre, 1932, p.240).

This quotation underscores his beliefs about choice between possibilities for human communities. Such a view coalesced as a possibilistic viewpoint, in contrast to the deterministic viewpoint presented by Ratzel in the first edition of *Anthropogeographie*. In the study of these human possibilities, he advocated a holistic approach: 'the geographer should study the whole and show that every region is a composite of mutually interdependent parts' (Buttimer, 1971, p.52).

Traceable throughout the above discussion of this genius's work are two major concepts: the *milieu* and the *genre de vie*. The final component in this conceptual trilogy was the concept of *circulation* which, he suggested, was central to the fostering of interaction between various parts of the world, especially the industrially developed world. This trilogy of concepts forms the foundation of his social geography.

Vidal's qualifications as a geographic exemplar are ably summarised by Buttimer:

> History acclaims Vidal for his prudent and stimulating insights rather than for any disciplinary formula, for his artistic finesse in geographical description, his power to suggest rather than convince, to evoke ideas rather than impose doctrine, good to open new horizons rather than define frontiers (Buttimer, 1971, p.58).

Possibilism, developed from the works of Vidal, continued to grow and spread on both sides of the Atlantic. In France, the major proponents and refiners included Jean Brunhes whose unfoundering support to the possibilist viewpoint and empiricism is well documented (see Buttimer, 1971, ch.4). It was Brunhes who enunciated the first explicit formulation of human geography as a systematic approach to the study of geography. So complete was the influence of Vidalian tradition (*la tradition vidalienne*) that the chairs in almost all of the 16 French universities, by 1921, were pupils of Vidal. Paraphrasing Joerg (1922, pp.438-41), James notes that 'in no other country . . . not even in Germany around Richthofen, has the development of geography been so centered around one outstanding teacher' (James, 1972, pp.249-50).

Outside France, the possibilist ideas, in contradistinction to environmental determinism, were accepted by a large group of geographers and anthropologists (see Grossman, 1977). In American geography, this influence was not pronounced until the 1920s, when Barrow's ecological conceptualisation of geography stressed the non-recursive feedback relationship between man and the environment. This ecological viewpoint was too circumscribed for many American geographers and it never really gained acceptance. A more acceptable and less circumscribed view of possibilism was presented by Sauer.

Sauer's Morphology of the Landscape. During the 1919-22 period Sauer's four papers before the Association ('Economic Problems of the Ozark Highlands of Missouri' in 1919, 'Geography as Regional Economics' in 1920, 'Problems of Land Classification' also in 1920, and 'Objectives of Geographic Study' in 1922) were regarded as significant extraordinary contributions to geography. As James and Martin noted, 'future chroniclers of the history of geographic thought may well view this contribution as Kuhnian "extraordinary science"' (James and Martin, 1979, p.71).

Sauer's ideas, in his earlier four papers, were more fully developed in 'The Morphology of Landscape' (1925). He rejected areal differentiation, asserting that the geographer's role is to investigate and understand the nature of the transition from the natural to the cultural landscape, and the successive stages through which the cultural landscape has passed during its transformation. From such an exercise the geographer would identify the major changes that have occurred in an area as a result of occupancy by a succession of human groups. As he puts it,

> our naively given section of reality, the landscape, is undergoing manifold change. This contact of man with his changeful home, as expressed through the cultural landscape, is our field of work. We are concerned with the importance of site to man, and also with his transformation of the site (Sauer, 1925, p.53).

Because of the difficulties in the analysis of the transition from the natural to the cultural landscapes, Sauer had to modify his original goal of understanding the transition from natural to cultural landscapes. He developed a historical framework for the study of landscape development with a focus on the patterns of human occupancy rather than on the socio-cultural agencies that generate the patterns. In his 'Foreword to Historical Geography' (1941), he fully developed this historical viewpoint; a view which permeated his subsequent works and those of his students (see Leighly, 1976, for a review of how Sauer developed this viewpoint). In such studies the emphases were on occupation and land use. His later works, including *Agricultural Origins and Dispersals* (1952) and *The Early Spanish Main* (1966), reflect this focus. 'The dimension of time', he later wrote, 'is and has been part of geographic understanding. Human geography considers man as a geographic agent, using and changing his environment in non-recurrent time according to his skills and wants' (Sauer, 1974, p.192).

Historically, Sauer was influenced by many European geographers, especially Otto Schlüter and Siegfried Passarge. Schlüter believed that the distinctiveness of a region could best be studied by a reconstruction of the original landscape, the *Urlandschaft*, followed by a historical account of the stages through which the landscape was transformed by man, the *Kulturlandschaft*. Slightly different from Schlüter, was the view of Passarge that the landscape was not unique, it was a type. As James noted, 'he saw a landscape type as what we would call a spatial system, an assemblage of interrelated elements' (James, 1972, p.235). In the discussion of the fourth dimension of geography, time, Sauer pays tribute to these people and other European geographers who directly influenced him:

> In 1923 I moved from Michigan to California to gain experience of a different country, and also to get away from what geographers mainly were doing in the East, which interested me less and less as narrowing professionalism. I had begun to read seriously what German, French, and English geographers were learning about the world as long and increasingly modified by man's activities. The *The Morphology of Landscape* was an early attempt to say what the common enterprise was in the European tradition (Sauer, 1974, p.191).

Sauer's contributions to the evolution of possibilism and American geography has been immense indeed. His leadership of the Berkeley School of Geography is a testimony to his teaching and research.

Hartshorne's Nature of Geography. Unlike the first three exemplars, the fourth was largely a synthesis of, and commentary on, a viewpoint that was evolving on both sides of the Atlantic. In Europe, the German School of Geography had moved from the earlier Ratzelian extreme determinism and developed the chorological viewpoint. Chorology as an approach to the study of geography was first developed by Richthofen and elaborated and refined by Alfred Hettner in a series of papers published in *Geographische Zeitschrift*. In America this viewpoint was proposed and defended by Hartshorne. As he pointed out:

> the goal of the chorological point of view is to know the character of regions and places through comprehension of the existence together and interrelations among the different realms of reality and their varied manifestations, and to comprehend the earth surface as a whole in its actual arrangement of continents, larger and smaller regions, and places (Hartshorne, 1961, p.13).

Hettner, like Hartshorne, argued that geography could use both the idiographic and nomological approaches; although many scholars, including Bunge (1962) and David Harvey (1969, p.50), argue that the use of place by Hartshorne, implied an idiographic approach. As Guelke recently noted, Hartshorne's fuzziness about the idiographic viewpoint, and his advocacy of the scientific approach to geography was, in part, responsible for the difficulty in classifying his position on the idiographic/nomothetic debate (Guelke, 1977). Although the works of Hettner, published in *Geographische Zeitschrift*, constituted the initial 'exemplary views' on chorology, and although this viewpoint was first directly articulated in the United States by Fenneman in 1919 (Fenneman, 1919), and many papers on the regional or area viewpoint appeared subsequently (see James and Martin, 1979, p.77, for examples), it was Richard Hartshorne's *The Nature of Geography* (1961) that codified this chorological viewpoint into one of areal differentiation. As Taaffe noted, 'Hartshorne used the term, areal differentiation to characterise the way in which geographers dealt with the wide variety of phenomena physical, economic, and social, which exist together in area and distinguish them from other areas' (Taaffe, 1974, p.6).

Although criticised over the years, *The Nature of Geography* was, and it is still, a central textbook in courses on geographic thought; it has also provided, for decades, research foci for geographers all over the world. In part, *The Nature of Geography* was a synthesis of the works of European and American geographers, but it also provided, by incisive comments and through subsequent elucidations in papers and conferences, a geographic research viewpoint that advocated regional studies emphasising areal differentiation. We share Guelke's recent comment about the importance of *The Nature of Geography*: 'If one seeks to understand recent developments in geography one must begin with an examination of *The Nature of Geography*. Most subsequent developments, I argue, were not revolutionary but involved the logical extension of key ideas to be found in that book' (Guelke, 1977, p.376).

Schaefer's Exceptionalism. The 1950s and 1960s were periods of discontentment with the subject matter of geography, and the widespread acceptance of quantitative methods made changes and shifts in our discipline inevitable. The background to these changes has been ably summarised by Taaffe:

> As the integrative studies of the forties and fifties proceeded, the absurdity of attempting genuinely holistic studies without clearly stated selection criteria soon became evident . . . Most geographers

were not really trying to synthesize everything in an area, nor were they trying to synthesize all phenomena of significance to man which had significant spatial expression. This increased awareness of a spatial bias in the selection of problems led in the fifties and sixties to increasingly explicit statements of the spatial view in geography (Taaffe, 1974, pp.6-7).

Historically, Schaefer's attack on exceptionalism in geography and his advocacy for a geography that is more nomothetic and based on spatial theories, may mark the inception of the spatial viewpoint in geography (Schaefer, 1953). Schaefer advocated the use of spatial laws as a basis for geographic explanation. He wrote:

description, even if followed by classification, does not explain the manner in which phenomena are distributed over the world. To explain the phenomena one has described means always to recognize them as instances of laws (Schaefer, 1953, p.227).

The spatial viewpoint advocated by Schaefer was restated by Ullman in the same year. He suggested that geography should focus on spatial interaction. 'The main contribution of the geographer', he noted, 'is his concern with space and spatial interaction' (Ullman, 1953, p.56).

The untimely death of Schaefer deprived the discipline of further insights into his thoughts about spatial laws and geographic explanation. Thus, unlike the other exemplars that benefited from further elaborations by the authors, the spatial viewpoint had to develop through the articulations of future scholars. Bunge's *Theoretical Geography* (1962), Haggett's *Locational Analysis in Human Geography* (1966), and David Harvey's *Explanation in Geography* (1969) are basically elaborations of Schaefer's original criticisms and suggested modifications to Hartshornian orthodoxy. Largely because of the concerted intellectual contribution to the spatial theme, the major tenets have gradually evolved; from Ullman's spatial interaction theme to the theme of spatial organisation which was institutionalised by Abler, Adams and Gould (1971). Subsumed under the umbrella of spatial organisation are numerous studies on spatial interaction, spatial systems, social planning and regional taxonomies. In all the associated studies, the emphasis is on both patterns and processes.

In the discussion of what we regard as exemplars in geography we have generally focused on those works that have had profound pedagogical and methodological import in the discipline. For a discipline that has evolved over a century, it is, of course, difficult to discuss the works of all those scholars that have had an impact upon the field. Here we were only interested in those works we regard as initiating significant changes in the focus of the field over an appreciable period of time.

Paradigms in Geography

The above discussion of exemplars suggests that paradigms may have developed around them. Here we explore this possibility.

As noted earlier in this chapter, a viewpoint may become a paradigm if, in addition to an exemplar, it had a large following and if it had clearly defined theoretical and methodological bases. In Table 1.1 we have summarised these for each of the exemplars

Table 1.1: Geographic Exemplars and their Associated Attributes

Author(s)	Exemplar	Image of Subject Matter	Theories & Laws	Methods	Area of Discontent
Ratzel	Anthropogeo-graphie	Inorganic control-Organic response	Darwinism(stages theories. of Physical and Social systems) Environmental Determinism	Deductive Newtonian cause & effect. Systematic approach	Too much environmental control. Absence of Human Initiatives & choice: eliminated novelty, & creativity
Vidal	Tableau de la Geographie de la France	Changes in attitudes values and habits create possibilities for human communities	Genres de vie: the product & reflections of the interrelation between men & his environment	Field work/ case studies Emphasis on causal successions or sequences.	Too much emphasis on a region (pays)
Sauer	The Four 1919-1922 Papers Morphology of Landscape Agricultural Origins & Dispersals The Early Spanish Main	Time the Fourth Dimension in Geography Landscape view of geography	*Cultural landscapes evolve from the physical landscape *Man as an agent of environment modification	Field work & historical re-construction of the human landscape Inductive development of landscape patterns	Pre-occupation with pattern rather than process The inability to analyse societal values, beliefs & social organi-sation.
Hartshorne	Nature of Geography Perspective on the Nature Geography	Chorology Perceived as idiographic	*Functional relation-ship *Order-classifications	Field work Mapping	No laws No generalisations Too restrictive a view
Schaefer	Exceptionalism in Geography	Spatial Interaction Spatial Organisation Nomothetic appeal	Location Flows Distribution Settlement	Mathematics Statistical methods Scientific Method	The method is too restric-tive and value-less.

*May not be theories or laws in the traditional sense.

discussed above. Based on these attributes and the fact that each exemplar had a significant following, we can tentatively assign paradigmatic status to them — Ratzel with the paradigm of determinism, Vidal with that of possibilism, Sauer with the landscape paradigm, Hartshorne with the chorological paradigm and Schaefer with the spatial organisation paradigm. In a historical framework, Ratzelian determinism was partly contemporaneous with Vidalian possibilism, Sauer's historico-anthropological paradigm was basically contemporaneous with Hartshornian chorology. The modifications of both chorology and the landscape view gave rise to the spatial organisational paradigm.

The above paradigmatic framework is only valid if we abandon the Popper-Kuhn view that a new paradigm is created as a result of a revolution. Indeed, Ratzelian determinism and Vidalian possibilism were very different and extensive debates did occur between their numerous disciples, but there was no real revolution prior to the emergence of possibilism. As noted earlier, when two paradigms co-exist simultaneously, the discipline is a dual-paradigmatic discipline. Similarly, the historico-anthropological and the areal differentiation viewpoints may have been divergent, but one did not emerge as a result of a revolution; a dual-paradigmatic discipline was still operational.

It should be noted that the dual-paradigmatic period we have associated with the Sauer-Hartshorne viewpoints does not imply that environmental determinism and possibilism, as per Ratzel and Vidal, were defunct. It implies, however, that their dominance as research and pedagogical vehicles had diminished. This tendency for older paradigms to persist beyond their dominant periods was particularly true of the 1960s and 1970s when the spatial paradigm was in vogue. Concomitantly, the historical approach of Sauer and the areal differentiation approach of Hartshorne were still operational. In spite of this, we believe that in the late sixties and early seventies geography was essentially a single paradigm discipline. New geographical journals such as *Geographical Analysis* (Golledge, 1979) and *Environment and Planning* became vehicles for theoretical work on this spatial theme. *Economic Geography* and *The Transactions* also changed to reflect this theme.

The spatial paradigm was capable of such general acceptance because it was initially packaged in the framework of a specific philosophy, logical positivism. Indeed, *Explanation in Geography* was a timely vehicle for linking the spatial paradigm to laws and theories through rigid mathematisation. The Resource Paper Series of the Association of American Geographers and the Concepts and Techniques in Modern Geography (CATMOG) Series of the Institute of British Geographers were suitable vehicles for conceptual, theoretical and methodological dissemination of the spatial viewpoint. These circumstances, coupled with the advancements in computers, created the 'bandwagon' atmosphere of the 1960s (Guelke, 1977, p.376). Books such as *Spatial Organization* (Abler *et al.*, 1971) and *Models in Geography* (Chorley and Haggett, 1967) were timely for the consolidation of this viewpoint, and the emergence of a single paradigm.

To summarise, if the revolutionary criterion for new paradigm development was relaxed, then geography had evolved from being a dual-paradigmatic discipline to a single-paradigmatic discipline. However, if the Kuhn-Popper criterion were maintained, only the spatial organisation paradigm may have evolved as a result of a revolution. To elaborate, Johnson (quoted by Harvey, 1972, p.3) has suggested that for a theory or paradigm to be accepted, it must possess certain basic characteristics: it had to attack the main theme of the prevailing orthodoxy; the theory must be new, yet it must incorporate some components from the existing orthodoxy; it must be relatively difficult to understand so that 'senior academic colleagues would find it neither easy nor worthwhile to study . . . At the same time the new theory had to appear both difficult enough to challenge the intellectual interest of younger colleagues and students'; offer a new methodology; and finally offer important empirical relationships for measurement and analysis. Of all the paradigms we have discussed only the spatial one seems to meet these criteria. On second thought, maybe there has been no paradigm change, only an extension. In this context, Gould's recent comment is appropriate:

> We see ourselves in an enlarged paradigm, and this is what I implied when I said that the dimensionality of geography has increased. We are not flip-flopping around from one paradigm to another, but enlarging our perspective by reaching out (Gould, 1979, p.145).

By the very nature of the exemplars discussed, geographic paradigms are examples of artefact paradigms. Since, as noted earlier, artefact paradigms may be based on artefacts such as a book, they may be sub-theory, and their development may not necessarily fit the Johnson desiderata of the Kuhn-Popper revolutionary process. Social sciences such as geography may change by extension and reorientation rather than by revolution. Consequently, we believe that our discussion of geography as being initially dual-paradigmatic and ultimately moving toward a single-paradigm science in the 1960s and 1970s may be valid. The major themes in geographic thought in the late 1970s and the trends in the 1980s are examined below.

Geographic Thought in the 1970s and 1980s

As noted in the last section, the spatial organisation paradigm was associated with a specific philosophy – positivism. As Guelke recently noted, 'the narrow philosophical choice offered to geographers in the realm of explanation made the wholesale adoption of a nomothetic approach all but inevitable. For many, geography was either science or mere description. If science, it followed that precision was important' (Guelke, 1977, pp.381-2). Burton's 1963 assertion that the quantitative revolution was complete (Burton, 1963), was a reflection of the global bandwagon effect noted earlier. From 1963 up to the beginning of the 1970s, the positivistic approach was virtually unchallenged, and David Harvey's *Explanation in Geography* was 'an attempt to establish the new geography on stronger philosophical foundations' (Guelke, 1977, p.382).

Although the spatial viewpoint was not directly challenged, the initial criticisms against this viewpoint were directed against its philosophical, methodological and theoretical bases. By its very nature, positivism is concerned with aggregate patterns, with the explanation and prediction of spatial patterns. In these ventures, man is portrayed as rational, and his spatial behaviours as reflections of an organism that follows spatial strategies which maximise some subjective utility function. Consequently, to quote from Bunting and Guelke, 'in developing theory human geographers, like economists, had relied heavily on a priori or normative models, which were based upon assumptions of perfectly rational people, isotropic planes and the like' (Bunting and Guelke, 1979, pp.449-50). In reaction to such a rational Cartesian system Buttimer wrote:

> The Cartesian social-scientist-cum-social-engineer perspective with its built-in values of efficiency, rationalization of agrarian structures, stream-lined transportation grids, and hierarchically ordered service networks, juxtaposes itself against my native world view with its own rationality,

its own 'ethnoscience' of the situation, construing the future (or
neglecting to think about it) in its own terms (Buttimer, 1974, p.2).

Similar views of the positivistic approach to explanation have been expressed
by numerous scholars including Guelke (1971) and Zelinsky (1975).

Criticisms of the existing spatial orthodoxy also grew from
unhappiness with the explanations provided for certain spatial
patterns. This criticism has been ably argued by David Harvey
(1972). He contends that when given the choice of alternate
explanations, the existing practice is to adopt that explanation which
tends to reinforce and, in many instances, legitimise the *status quo*.
Elsewhere (Harvey, 1974) he demonstrated, from his discussion of
population and resources, how the value system of the researcher
affects the types of conclusions. This is a direct attack on the value-
free explanatory approach advocated by the positivists.

Because of criticisms like those discussed above, the late 1970s
were characterised by (a) re-evaluations and attempts at modifying
the prevailing viewpoints to accommodate these criticisms, and
(b) the search for new philosophical viewpoints that may provide
new insights into spatial explanations.

a. Modifications of Existing Orthodoxies. In his paper 'Alternatives
to a Positive Economic Geography' (King, 1976), King examines
the possibilities of a social science that 'considers values along with
facts, and that allows for applied social science to contribute to the
development of policy paradigms' (King, 1976, p.306). He suggests
that in the analysis of any system, the consequences of alternative
end values must be explored and evaluated. The researcher must be
aware of alternative goals, he must critically assess those goals and
be capable of communicating with policy-makers using scenarios
(what he calls the 'story-telling based on metaphors'). King's
concluding statement is a plea for such modifications in the existing
orthodoxies. 'We should lower our mathematical sights and aim at
the target of developing operationally useful models rather than at that
of formally proving existence theorems and the like' (King, 1976,
p.308).

From a different perspective, Gould feels that the major
developments in geography, and the respectability that geography
has among social sciences, are due to the elucidating works by
mathematically talented geographers: Tobler's work has led to 'the
development of true cartographic research, genuine intellectual inquiry

that has enlarged our definition and our understanding of maps and the act of mapping' (Gould, 1979, p.147). Dacey's works have extended geographic theories and made contributions to mathematics, while Wilson's 'work in entropy maximization models has changed the way we look at the world . . . [and] have operationalised deep areas of statistical mechanics for geographic understanding' (Gould, 1979, pp.148-9). He commends Webber's ability to translate 'geographic observation and hypothesis from the verbal to the more manipulatable language of mathematics' (Gould, 1979, p.148). To Gould, modifications in the existing orthodoxy are possible when a 'more powerful ordering construct comes along' (Gould, 1979, p.145).

b. Adoption of New Philosophical Approach. Because of perceived weaknesses in positivism, many geographers have suggested alternate philosophical approaches to the study and understanding of spatial systems. In subsequent chapters of this book, the major tenets of these philosophies are presented, and their potentials for geographic analysis are evaluated.

Naturally, Chapter 2 presents a positivistic position of laws, objectivity, mathematisation and value-free analysis. The responses of Hill's ideal positivist to his ten-point self-assessment scale clearly presents the viewpoints of this philosophy. In Chapter 3, the pragmatist's viewpoints are presented. Like positivism, pragmatism advocates the use of the scientific method but for finding solutions to human problems. In the light of King's suggestions discussed in the last section, pragmatism may provide a suitable philosophical viewpoint in the 1980s for those who would like to modify positivism and use value-based scientific methodology to solve human problems. To the pragmatist, space is functional; the meaning of space is a function of the practical consequences of that space. Space is a composite of error and knowledge.

The third philosophical viewpoint that is discussed is basically a formalisation of a research focus that is common in geography: the study of functions and the application of systems theory. Functionalism is a viewpoint that investigates functional linkages with emphasis on the goals, the needs and the links between role and actors. The actual and potential applications of functionalism in geography are explored.

The first three philosophical viewpoints — positivism, pragmatism and functionalism — do not explicitly reject mathematisation; most of the other philosophies discussed in this book do. Major differences however exist among them in terms of the relevance of values in space, and the role of man in spatial analysis. Relph presents the phenomenological viewpoint. In this philosophy the emphasis is on human experience in space, the concept of geographicality captures the bonds that bind people to their surrounding, it manifests itself 'in a sense of place and in sensitivity' to landscape. The phenomenological viewpoint was institutionalised in the discipline in 1974 with the publication of Buttimer's *Values in Geography*. As the references in Relph's chapter indicate, it has slowly gained momentum and has become a credible alternative to positivism.

Closely related to phenomenology is existentialism. In existential geography, a central concept is that of existential space. Briefly defined, it is what Samuels calls the 'assignment of place'. Such an assignment is a result of human reality. Existential space, like spatiality, has two components — distance and relations. Related to these two components are the concepts of *partial space* which is the set of relationships linking man to his world, and the *situations of reference* which constitute the historical milieu in which the assignment of place is effected. Around these concepts, Samuels develops his ideas of an existential geography.

In contrast to the lived or existential space of existentialism and phenomenology, is the world view of the idealist. As Guelke had argued for some time, the world view of idealism is based on systems of ideas, and to explain any geographic phenomenon, one has to put that human activity into its cultural content. It entails a rethinking or reconstruction of that human activity in order to discover what really happened. For the idealist the mode of explanation is called *Verstehen* and Guelke has shown how it is an alternate method of explanation to the theoretical nomothetic approach of the positivist.

Antithetical in many respects to idealism is the philosophy of realism. Gibson suggests realism as a viable alternative explanation to idealism. To the realist, facts speak for themselves and explanation is logical and inductive. Unlike idealism, realism advocates the use of theories and models in geographic explanation. Although similar in the method of explanation to positivism, it is different in terms of the question being investigated. While the positivists focus on the 'how', the realists are interested in the 'why'.

Two other viewpoints that emerged in the 1970s and will continue to be important in the 1980s are Environmental Causation and Dialectical Materialism or Marxism. The rebirth of environmental causation (See Chappell's chapter in this book) is essentially a modern-day version of the paradigm of determinism discussed earlier in the chapter. Recent publications about the scarcity of global resources, and about the limits of growth in human societies as a result of resources and/or environmental constraints, makes a reappraisal of environmental causation timely.

Of all the alternative philosophical viewpoints to positivism, dialectical materialism seems to be the squeaky wheel. The chapter by Peet and Lyons does present the major tenets of this viewpoint. To the Marxist geographer, territorial structures essentially reflect the prevailing socio-spatial dialectics. Marxist geography analyses the dialectical relationships between social processes, the natural environment and spatial relations.

The above brief discussion of the alternative philosophical viewpoints to positivism was not designed to present encapsulations of these philosophies, but to give the reader an idea of their divergent conceptualisations of space; to underscore the fact that the spatial organisational paradigm is still in operation. The philosophical viewpoints we have discussed are just alternative ways (to some individuals, better ways) of investigating space and spatial relations. In the 1980s, as in the late 1970s, these philosophical diversities will constitute the major themes in geographic thought.

Conclusions

This chapter clarified the relationship between philosophy, thought and paradigm. Like the double-faced Janus, we then explored the development of geography as a paradigmatic discipline. This created a proper base for discussing the present and future themes in geographic thought. While philosophical diatribes may have been characteristic of the pre-spatial organisation paradigm, the large diversity of philosophical viewpoints and the messianic way in which they are presented are unique of the 1970s. We believe that the spatial paradigm will continue, and that geographic thought will increasingly focus on the diversity of viewpoints among the geographic philosophies.

2 POSITIVISM: A 'HIDDEN' PHILOSOPHY IN GEOGRAPHY

Michael R. Hill

> Twice or thrice had I loved thee
> Before I knew thy face or name
> John Donne

Introduction

As I write this, I am assuming that the reader will be a young geographer-in-training. This chapter is designed to help you 'uncover' positivism for yourself. I believe that philosophy is an activity rather than a set of dusty principles to be studied and memorised. The questions and exercises in this essay are not rhetorical. They are intended for your active response. Together, we will examine the general relevance of philosophy for geographic research, try to suggest why positivism has only recently become a frequently used word in geography and attempt to assess the extent of your sympathy for the positivist perspective. Later in the chapter, I will construct an 'ideal' positivist who will assert several ideas and principles. You will be invited to argue with the 'ideal' positivist, not just read what he has to say. Finally, I will present a list of suggestions for your own exploration of positivism and philosophy in geography, for not only is philosophy an activity, it is a personal activity. Only you can provide the philosophical answers that will have meaning and lasting value in your future work as a geographer.

Is Positivism Important?

Many geographers doubt that philosophical issues are actually relevant to geographic research. A partial answer to this question is found in the observation that no research (geographic or otherwise) takes place in a philosophical vacuum. Even if it is not explicitly articulated, all research is guided by a set of philosophical beliefs. These beliefs influence or motivate the selection of topics for research, the selection of methods for research, and the manner in which completed research projects are subjected to evaluation. In short, philosophical issues permeate every research decision in geography. The philosophical perspective called positivism is one set of beliefs (among many possible sets) which outlines what should be studied, how it should be studied, and the significance that should be attached to the findings. Positivism (in particular) and philosophy (in general) are personally relevant to students who want to have an explicit and systematic understanding of the set of beliefs which will guide them in their future work as geographers.

Opening the Door to Philosophy

Despite the above assertion that philosophy is important, it is easy to find geographers who dismiss philosophy as generally irrelevant. These naive individuals generally miss the whole meta-disciplinary focus of philosophy proper and usually conceive 'philosophy' in geography as concerned with the history of intradisciplinary debates over the 'definition' of geography. This situation is so serious that the remainder of this section will be devoted to the argument that the young geographer-in-training must become open to the discussion of truly philosophical issues in geography.

Philosophical Issues in 'Practical' Geography

Many supposedly down-to-earth, practical geographers point out that they are busy doing geography, are interested only in the objective study of geographic problems and are pretty much bored by the question: 'What is geography?' This position is unsatisfactory since several inherently philosophical problems spring immediately to mind: How is a 'problem' defined? How does one decide what methods are best for attacking the 'problem'? If one is bold enough to ask these questions, it may be suspected by the hard-nosed, practical geographers that there is something lacking in your geographic education. They will feel that the answers should be obvious to the student of geography. But there is nothing at all 'obvious' about the answers!

The 'Knitting Together' Process. To grasp why the answers may 'seem' obvious, but are not, one needs to reflect on the content of a typical geographic curriculum. A typical programme of study undoubtedly includes the following: attending lectures, completing exercises, preparing for exams, writing seminar papers, and studying examples of geographic research in textbooks. These activities are designed explicitly to teach you the facts, techniques, and concepts of geography. In essence, you are learning what Thomas Kuhn (1970a) would call the paradigm of geography.

Undoubtedly, the facts, techniques and concepts of geography are not isolated. They are 'knitted together' to form a conceptual network of interrelated ideas. Such a network is the 'paradigm' of a discipline. For example, the techniques of measuring area and counting individuals are utilised together with the 'concept' of density in arriving at the 'fact' that the population density of a given region is so many persons per square mile. During the course of study the student comes to 'knit' many such techniques, concepts and facts together. The practical, hard-nosed geographers expect that you will 'see for yourself' how all of these aspects of geography 'knit together' to form the core of the geographical perspective. However, this process does not involve seeing anything nearly so much as it involves the reduction of cognitive dissonance.

Through the 'knitting together' process, the student learned what topics and methods 'fit' within the discipline of geography as a whole. You have learned what things to expect when you open a geography textbook. As your training advances, there are fewer and fewer surprises when you read the latest issue of professional journals of geography. As his grasp of the nexus of ideas that constitutes geography grows more comprehensive and more integrated, it will appear more and more 'obvious' why a particular researcher works on a given topic.

You would probably not be surprised by a classmate's decision to do a paper on regional variations in population density because it is an 'obvious' topic within the accepted conceptual nexus that forms geography. Yet why should this topic be included in the geographic enterprise? To answer only that it fits within the cluster of accepted geographic ideas is an appeal to tradition and authority. Why should you be bound by this tradition in the choice of research topics and research methods? In asking this question, you have opened the door for philosophic discussion.

The Purposes of Geography and Their Implications. Your question might be answered by the down-to-earth geographers with the assertion that the topic of regional variations in population density, for example, is traditionally included as a topic in geography because it fits within the overall purpose of geography. This response shifts the discussion to an entirely different level. It is common to hear that the 'purpose' of geography include at least the following:

(1) To further man's knowledge.
(2) To help solve major social problems.

(3) To gain a conceptual understanding of man's organisation of space.

You have probably already given silent assent to one or more of these expressed 'purposes' of geography. But, have you ever thought about the questions which such assent implies? Each of the above stated purposes of geography suggests a number of questions. The questions are deeply philosophical, but you will find that you will be able to answer most of them briefly and quickly. For example:

(1) What is the nature of knowledge?
(2) How is knowledge extended or furthered?
(3) What is the benefit of extending knowledge?
(4) When is a situation identified as a social problem?
(5) What things does one need to know to solve a social problem?
(6) What is the geographer's proper role in the solution of social problems?
(7) What constitutes conceptual understanding?
(8) What is space?
(9) What is organisation?

Take a pencil and jot down a short reply to each of the above questions. Then, take a few moments to relax and reflect. Review each of your answers and ask yourself: 'On what did I base my responses?' Take an additional few minutes to outline the set of beliefs on which your answers were based.

It is significant if you were able to supply short definitions or give relevant examples in response to the above questions. Your ability to respond demonstrates that you already have a philosophical foundation or groundplan on which to erect answers to deeply philosophical questions. Yet you probably had more difficulty when asked to outline the basis (the philosophical foundation) of your responses. How can this be? Like the hard-nosed, practical geographers who find philosophy largely irrelevant, you may have formed a philosophical orientation without much conscious reflection. The key to understanding how this may have happened lies in understanding the implicit learning which accompanies the 'knitting together' process. In learning the concepts, techniques, facts and assorted purposes of geography, you had to find a way of fitting them together which made sense to you. While your mind was consciously occupied with specific concepts, techniques and facts, you were also internalising, developing and reinforcing the basis on which you knitted together these explicit aspects of geography. This is a very natural tendency. It is similar to the way in which you were socialised into the culture of your native land. In other words, your philosophy of geography may have been learned implicitly while you were explicitly attending to the substantive contents of your courses, textbooks and assigned exercises.

The Relevance of Philosophical Reflection. You are now in a position to begin to appreciate the strategic relevance of philosophy for your future career as a geographer. The philosophical groundplan which you are now developing and internalising is important in several ways:

(1) Your philosophical foundation will guide you in the selection of research methods and topics during the remainder of your professional activity as a geographer. Should something so fundamental not deserve your conscious reflection and assent?

(2) Unlike your socialisation into a culture as a youngster, becoming a geographer is a choice which you have made. This means that you have some degree of intellectual and moral responsibility for the kind of geographic orientation to which you will ultimately give your allegiance.

(3) You are still relatively young and still have a chance to reflect on your philosophical foundations before they become firmly fixed. It is very difficult (intellectually painful, if you will) to call your philosophical beliefs into question once you are very far into your professional career. The best time for philosophical reflection is now.

If you see any merit in the propositions above, then you will not ignore philosophy as irrelevant and unrelated to the day-to-day activities of the geographer. That philosophy is not seen as relevant by many practicing geographers does not mean that their work is not guided by philosophical presuppositions. Philosophy is all around you. With this in mind, you are now prepared to continue with the central theme of this chapter: 'positivism'.

Becoming Aware of Positivism. Positivism is one of the unrecognised, 'hidden' philosophical perspectives which guides the work of many geographers. It is a perspective which has allowed contemporary geographers to 'knit together' many new techniques and concepts.

If you have been trained under the tutelage of professors who have adopted positivism as a perspective, there is a good chance that your developing philosophical groundplan already incorporates several 'positivist' ideas, ideas which have been passed on to you for your unreflective, uncritical acceptance. When you have some notion of the outlines of positivism, you may want to reject it in favour of some alternative philosophical framework. On the other hand, you may discover (like the poet quoted at the beginning of this chapter) that you always liked the ideas of positivism, but you just did not know its full implications or proper philosophical name.

Are you a Positivist?

Positivism may be an unfamiliar word to you, but even as a neophyte geographer you are probably not unfamiliar with many of the central beliefs of this philosophy. Although this may surprise you, many of the fundamental tenets of positivism are examplified in virtually any current issue of the mainstream geographic journals and in many recent textbooks. Haggett *et al.* (1977, p.23) note that positivism has 'played a major part in geographical explanations of phenomena' in much geographic work of the past. Paradoxically, positivism remains as a hidden philosophical perspective in the sense that those who adhere to many of its central tenets rarely describe themselves as positivists. This is unique in geography today. Those who hold to other philosophical systems routinely label themselves, for example, as 'idealist', 'humanist', 'structuralist', 'existentialist', 'Marxist', 'phenomenologist' and so on. While many boldly carry the banner of their chosen philosophy, the name of positivism is rarely seen or heard in the works of geographers who give assent to its basic principles. We will explore some hypotheses for this curious situation in the next section. In the meantime, it is important for you to realise that you have undoubtedly been reading and studying positivist geography at one time or another, but you probably did not recognise it as such. In fact, you may have a better implicit understanding of positivism than any other philosophical perspective in geography. Finding out the extent to which you have already given silent assent to the principles of positivism is the purpose of the exercise in Table 2.1.

Having computed your 'PS', you now have some idea of the extent to which you may already by leaning toward the positivist perspective. It should provide good class discussion if you compare

Table 2.1: The Ten-Point Self-Assessment Positivism Test

Instructions: Carefully consider each statement below. Mark the response which most closely resembles your true beliefs. The possible responses are purposely dichotomous to keep you from sitting on the fence

/1/   Geographers should be very careful not to allow political or religious beliefs to influence their interpretation of geographic data.

/ / Agree   / / Disagree

/2/   Geographers can legitimately claim that they use a research method which distinguishes them from all other scientists.

/ / Agree   / / Disagree

/3/   An 'hypothesis' and a 'theory' are essentially the same thing.

/ / Agree   / / Disagree

/4/   The subject matters of human and physical geography are so inherently different that it is impossible deductively to link human and physical geography together in a coherent, logically unified body of theory.

/ / Agree   / / Disagree

/5/   All propositions in geography should ultimately be expressed in the language of symbolic logic.

/ / Agree   / / Disagree

/6/   Although a geographic theory may account for present spatial patterns, it is unreasonable to insist that it should also account for past patterns as well as future patterns.

/ / Agree   / / Disagree

/7/   Explaining an event scientifically involves little more than showing that the event is statistically predictable.

/ / Agree   / / Disagree

/8/   It is possible to prove empirically that some theories are false and that other theories are true.

/ / Agree   / / Disagree

/9/   Any empirical assertion made by a geographer which cannot be empirically verified by other geographers is not a particularly useful assertion.

/ / Agree   / / Disagree

/10/ Because scientists rely on physical observations for their data, they are never justified in making references to theoretical entities which cannot be physically observed or measured.

/ / Agree   / / Disagree

Scoring Instructions: Compute your 'PS' (Positivism Score) by using the key in Table 2.2.

Table 2.2: Scoring Instructions for the Positivism Test

Score one point for each 'correct' response. If you are an 'ideal' positivist, you will receive a score of '10'.

/1/	Agree	/6/	Disagree
/2/	Disagree	/7/	Disagree
/3/	Disagree	/8/	Disagree
/4/	Disagree	/9/	Agree
/5/	Agree	/10/	Disagree

---

your score with those of your classmates and try to account for the differences in your scores. The rationales that would lead a dyed-in-the-wool positivist to obtain a perfect score of '10' will be presented later in this chapter. First, however, if you have discovered that you already have some positivist leanings, you may be wondering why it is that 'positivism' is not a frequently used term in mainstream geography.

Uncovering Positivism

In this section, three hypotheses are proffered as possible explanations for the infrequent use of the term positivism in contemporary geography. Let us review them briefly:

(1) The Self-Evidence Hypothesis: Positivism is so well-known and so thoroughly understood that even the neophyte geographer can be expected to know when an author is writing from the viewpoint of positivism. It is really unnecessary for an author to indicate that he/she is a positivist because it is self-evident that such is the case.

(2) The Sub-School Hypothesis: Positivism is a very broad term which covers a wide range of different but related philosophical perspectives or sub-schools. It is inappropriate for an author to indicate that he/she is a positivist because it would be technically misleading. Rather than call himself a positivist, an author is expected to identify the sub-school to which he/she subscribes.

(3) The Trojan Horse Hypothesis: Philosophical perspectives which assist in knitting together the concepts, techniques and facts of other disciplines have sometimes been imported into geography

without conscious reflection. A geographer who explicitly borrows concepts, techniques and facts from a positivist cognate discipline may also implicitly borrow a positivist outlook. An author may thus be unfamiliar with the formal name of this perspective.

Hopefully, you will find a germ of insight in each of these propositions. No single hypothesis should ever be expected to account for everything.

The Self-Evidence Hypothesis

Strictly speaking, the first hypothesis would not take you very far in understanding the hidden nature of positivism in geography. It is doubtful that very many geographers (especially neophyte geographers) could recite the canons of classical positivism. Viewed in historical context, however, this hypothesis has much to offer.

It is important historically to realise that a great many of the philosophy of science texts and courses available were written and taught by philosophers rooted (at some time) in the positivist tradition of philosophy. In addition, many philosophers who are no longer positivists in a strict sense have taken positivism as a philosophical starting point so that many of the issues which they address derive from the positivist tradition. In short, positivism has been a dominant influence in setting problems and thereby shaping the evolution of contemporary philosophy of science.

This historical note is important for geography to the extent that many geographers have endeavoured to make geographic work more scientific. When geographers have turned to philosophers of science for help in this enterprise, they have (virtually by definition) frequently turned to philosophers either grounded in positivism or attempting to solve problems which positivism brought to their attention. As a result, the first hypothesis makes considerable sense if you are willing to perform a slight lexical modification by substituting the term science for the term positivism. It is not at all difficult to find geographers who can give a reasonable account of what it means to be scientific. Further, many authors who proceed in a scientific manner often assume that it is self-evident when they are doing so.

The substitution of one term for another (together with the historical tradition of philosophy of science in the United States) has had, in geography, an unfortunate consequence which is not always appreciated by geographers of the scientific turn. To wit, there has been a tendency on the part of scientific geographers to view the philosophy of science with which they are most familiar (a philosophy of science with positivist roots) as the philosophy of science. This has resulted in stifling philosophical discussion in many quarters of geography. Alternative philosophies of science (i.e., non-positivist and anti-positivist viewpoints on the nature of science) have not been judged on their merits as philosophies but judged against the requirements of science as derived from positivism. Given the historical development of the philosophy of science, geographers of the scientific school should not be reprimanded too harshly for their past provincialism. But, ever since the publication of Buttimer's (1974) sensitive essay on the nature of values in geography, there has been no justifiable defence available to geographers who fail to adopt a more catholic attitude towards science.

The Sub-School Hypothesis

The sub-school hypothesis helps us to understand how it is possible that geographers who studied positivist philosophy of science did not pick upon the term positivism. What is referred to today as positivism (or logical positivism or logical empiricism) emerged in the 1920s and was developed by an impressive group of philosopher-scientists who banded together to form what came to be known as the Vienna Circle. After many meetings and discussions, these scholars published a manifesto in 1929 which outlined their basic philosophical programme for the reform of science (cf. Joergensen, 1951). Although there was general agreement that science was in sore need of reform, the men who became members of the Vienna Circle were not all of like mind when it came to discussion of the technical details of their general proposal. Subsequent debates led eventually to development of identifiable substreams or sub-schools of thought within the general positivist movement. (Discussion of the characteristics of these separate sub-schools lies outside the scope of this chapter. An outline of their main features is found in Radnitzky (1970).) In short, the positivist movement was never a really unified school of thought. The members of the Vienna Circle did not apply the term positivist to themselves. It was applied by others as an easy way to identify a cluster of related philosophical beliefs.

Geographers who delve into the philosophical works of positivist philosophers may not encounter the term positivism with any great frequency. They are most likely to find a version of the original programme of the Vienna Circle together with a specific commentary on the manner in which particular points in this general programme should be realised. The same is true in many courses on the philosophy of science. The issues raised by the members of the Vienna Circle tend to be taken as 'givens' and class time is often devoted to discussion of the ways in which these issues can be solved by the practising scientist. Unless you enroll specifically in a course on the history of contemporary philosophy of science, it is possible that you could be indoctrinated into the basic viewpoint of the Vienna Circle without ever coming into contact with the term positivism. The sub-school hypothesis also relates to the works of geographers who exhibit above-average (for geographers) philosophical sophistication. They tend to identify with and discuss particular points of view within the framework of contemporary philosophy of science rather than with a grand (and now outdated) generalisation such as positivism.

The Trojan Horse Hypothesis

The Trojan Horse hypothesis has implications which are less flattering for geographers. Almost every geography department has one or two geographers who want to be scientific (usually for reasons of prestige) without doing their philosophical homework. Rather than work to understand what is implied philosophically in scientific procedures (to the extent that they could develop concepts and techniques of their own design), they borrow concepts, techniques and facts from those disciplines which are generally recognised by the academic community as being more scientific than geography. This opens the door for unreflective development of positivistic tendencies within geography.

As the borrowing continues and the borrowed items are made to fit within the conceptual nexus of geography, geography becomes more and more like the disciplines from which the items were originally borrowed. If the disciplines from which the items were borrowed became recognised as scientific by virtue of their allegiance to positivist principles, then the items borrowed are undoubtedly compatible with the positivist perspective. As the borrowed items are made to fit within geography, the core of geography is modified so as to, at least, be compatible with the philosophy of positivism. Through this knitting together process, geography comes to include ideas compatible with positivism even though the geographers who did the interdisciplinary borrowing were never explicitly concerned with philosophical issues *per se*. Thus positivist ideas can slip into geography just as the Greeks slipped into Troy via the belly of a wooden horse. But unlike the Trojans who lost the battle, the geographers who have gone a-borrowing have not the slightest awareness that they have been conquered by a foreign power. Since their debt to positivism has not been developed as a result of conscious reflection, it is better to call their ideas positivistic rather than positivist. In the same vein, it is more appropriate to call their methodological procedures scientistic rather than scientific. Fortunately, the Trojan Horse hypothesis today describes fewer and fewer geographers.

Testing the Hypotheses on Your Own

I have found that the three hypotheses above help me to understand how it is possible to find positivist principles espoused by geographers who do not label themselves as positivists. Yet I do not expect you to accept these hypotheses without investigation on your own. Let me suggest that you engage your professors and/or associates in conversation concerning the role and nature of philosophy in geography. Ask them about their viewpoints concerning the relevance of science, values, logic, theory construction, the unity of science and philosophy in geography. Their answers will help you assess the utility of the hypotheses I have offered for your review. You may find that other hypotheses of your own formulation are more meaningful. But I am forging ahead of the game. Before engaging in discussion, you will want to have a general outline of some of the major principles of positivist philosophy.

Constructing an Ideal Positivist

Now that: (1) the extent to which you may be in sympathy with positivism has been assessed, and (2) some hypotheses to account for the general absence (until recently, at least) of the term positivism in the geographic literature have been suggested, it is time to outline the nature of positivism as a philosophy. This outline will allow you to compare your own beliefs more carefully with those attributed to positivism and to engage your professors in conversations designed to test the hypotheses offered in the preceding section. The outline of positivist thought will take the form of a set of expanded responses that might be expected from an ideal positivist to each of the questions on the Positivism Test. There are two major reasons why I am resorting to the use of an ideal type:

(1) As noted earlier, the positivist movement in philosophy was united for only a short period in the 1920s. Even so, the initial unity

displayed was relatively superficial. There was general agreement on what positivism ought to accomplish, but there was never unified agreement on the particulars concerning how the goals of positivism should be effected. These disagreements resulted in the sub-schools already noted. Thus the responses outlined in the next section are those of a fictional positivist or what the sociologist would call an ideal type. The responses pick up on themes discussed in a number of the sub-schools of positivism.

(2) Not only was there disagreement among the founders of the positivist tradition, the nature and focus of this disagreement shifted as the positivist perspective evolved. Positivism was never a set of iron-clad principles. It was always in flux. Thus there is a problem in deciding what era or temporal aspect of positivism to emphasise in fabricating the responses of my ideal positivist. I have generally resolved the issue in favour of early positivism because it is this version which has been most influential in the social sciences. Radnitzky (1970, vol.I, p.71) has pointed out that social scientists (and I view geography primarily as a social science) have failed to take account of more recent philosophical developments which grew out of the early positivist positions espoused by the members of the Vienna Circle: 'Instead they have remained users of the oldest, now very old-fashioned version of it. This is a striking example of cultural after-lag' (Radnitzky, 1970, vol.I, p.71).

If you are interested in an understanding of the influence of positivism in geography, a grasp of this older version of positivist principles will be very useful. In addition, you will be in a better position to appreciate the historical context of more recent developments in the philosophy of science. It is also worth noting that those who oppose positivistic thinking in geography may have this older version of positivism in mind when they level their critiques. If you are going to reject positivistically-rooted philosophy *per se* (as opposed to the influence of early positivism in geography), then you are well advised to consult more recent texts and articles in the philosophy of science before launching your attack.

The Responses of an Ideal Positivist

In this section I have fabricated the responses that my ideal positivist might be expected to give to each of the questions on the Positivism Test. Each response is followed by a brief commentary. The responses

and commentaries provide an outline of the beliefs commonly attributed
to positivist thinkers in the social sciences. I have written the responses
of my 'ideal' positivist in the first person in order to make your reading
a bit more lively. I admonish you not to assume that I necessarily agree
with the responses as given. Undoubtedly, you will not find yourself
in full agreement either. You may also find that you gave the correct
response on the Positivism Test for quite different reasons than those
offered by the ideal positivist.

If you are alone in your room reading this chapter on a Saturday
night when others are engaged in less scholarly pursuits, I invite you
verbally to challenge each response. This will sharpen your understanding
of your own philosophical beliefs. It is recommended, however, that you
read this section with at least one other classmate. One of you should
play the role of the positivist. The other should assume the role of
challenger. By arguing, you can better come to appreciate the nature
of the positivist viewpoint and decide for yourselves what constitute its
strong points and its weak points. Now, the responses:

/1/ *I agree.* Geography must be objective. If we allow personal religious
and political beliefs to influence our interpretation of geographic data,
we must grant the validity of metaphysical assertions which can never be
objectively verified in any way whatsoever. You undoubtedly realise
this, but previous geographers have not been so careful. For example,
Carl Ritter interpreted geographical patterns as evidence for the
handiwork and almighty wisdom of God. Yet he was never able to
present an objective rationalisation for his *teleological* interpretations.
The result was an intellectually short-circuited system of geographic
beliefs. A more sinister case is found in Karl Haushofer's *Geopolitic.*
This system of ideas was not designed to advance the discovery of
objective knowledge, but to legitimise the territorial demands of Nazi
Germany. Both Ritter and Haushofer serve as warnings of a serious
intellectual mistake, attending to metaphysical questions instead of
the search for objective knowledge.

Ridding philosophy of its preoccupation with metaphysical issues
was one of the most important tasks undertaken by myself and the
other members of the Vienna Circle. We worked very hard to devise a
method of philosophical and scientific investigation which would help
scientists like yourself avoid the metaphysical traps of religion and
and politics. Ludwig Wittgenstein, who met occasionally with the
Vienna Circle, was very helpful in showing us exactly why
metaphysical questions could not be answered by either science or

philosophy. Wittgenstein demonstrated that metaphysical assertions are linguistic traps. They involve us in making assertions about things we can never know in any objective way. He explained that it is not philosophically legitimate to speak about things which cannot be pointed to in experience. He concluded that 'What we cannot speak about we must pass over in silence' (Wittgenstein, 1963, p.151). Although Wittgenstein later changed his view on what was involved in attributing meaning to linguistic propositions, his early ideas influenced our zealous declaration that scientists and philosophers must not be tempted to seek answers to metaphysical questions. Our intent, however, was not merely to be destructive and negative toward previous philosophy. We replaced an old philosophy riddled with unwarranted metaphysical language with a positive contribution. We outlined the boundaries of science and gave scientists a method and a language for insuring that what they said or asserted would be objective and free of metaphysical presuppositions and assertions.

/2/ *I do not agree.* The methods of scientific investigation that we developed in our Vienna meetings are available to all scientists. We did not reserve any distinctive methods for particular areas of scientific inquiry. Your geographic forefathers, however, were not always clear on this point. For example, James and Jones (1954, p.16) claimed that geography is an identifiable discipline due, in part, to its possession of a 'distinctive methodology'. Rather than recognise that geographers were simply using methods employed by all scientists, James and Jones (1954, p.7) suggested that anyone who used the 'geographic method' was really a geographer at heart:

> Scholars who are not identified professionally as geographers are not the only ones who make use of the geographic method or who study and write *what is, in fact, geography.* Geographic writing is done by anthropologists, economists, sociologists, political scientists, botanists, zoologists, geologists, business men and many others. (Italics mine)

What a marvellous (and incredible!) sleight of hand. It is not obvious that the widespread use of the methods employed by geographers makes them anything but distinctive of geography? All scientists draw from the same armoury of methodological weapons. You are free to use any research methodology as long as it conforms to the canons of good science. Why restrict yourself?

/3/ *I do not agree.* The terms 'hypothesis' and 'theory' are not equivalent. The former is subordinate to the latter in the following way. An hypothesis is an empirically unverified assertion which has been derived from a more complex structure called a theory. In science, hypotheses flow from theories. An *ad hoc* hypothesis not derived from theory has no role in genuine science.

You must first grasp the nature of theory if you want to understand how hypotheses are produced. A scientific theory has several essential components:

(1) A logical superstructure of metatheory which specifies the rules of logic that will be used in the deductive expansion of the theory proper.

According to Woodger (1939, p.65), a theory is considered formalised 'when its metatheory is completely and explicitly stated'. All scientific theories share the same metatheory.

(2) A set of fundamental assertions or *axioms* which are taken as given.
(3) A set of *definitions* which define the terms (or *primatives*) used in the axioms.

You should note that primative terms in a theory can never be completely defined. We would become involved in infinite regress if we attempted to define them fully. A theory is considered *axiomatised* when it possesses a set of definitions and axioms from which all other concepts and assertions in the theory can be deduced as consequences (Woodger, 1939, p.66).

(4) A set of *theorems* or assertions which are subsequently formed by combining the definitions and axioms according to the logical rules in the metatheory.

It is important to realise that theories with these components make no empirical assertions of any kind whatsoever. They are completely abstract. They are useful to the scientist only when they are interpreted.

The process of interpretation brings me directly to the production of hypotheses. An hypothesis is simply an interpreted theorem. The procedure for interpreting a theorem is set forth in an additional theory component which is found in all scientific theories.

(5) A *text* which tells how a theory should be empirically interpreted.

For example, suppose that we have derived the Pythagorean theorem from the axioms of Euclidean geometry. So far, this is just an abstract theorem which is the deductive consequence of an uninterpreted theory. Before the theorem can be interpreted, we must know several things, including:

(a) The rules for empirically constructing a right triangle; and
(b) The rules of measurement so that we can make empirical observations on the right triangle we construct.

These two elements must be included in any text for Euclidean geometry if that theoretical system is going to have any empirical significance. With our text in hand, we can now form an hypothesis based on the Pythagorean theorem, for example:

If we have a right triangle with legs of 3″ and 4″ respectively, then the hypotenuse of this triangle should be 5″ (if the Pythagorean theorem is empirically meaningful).

We now carry out an empirical investigation. Using the text as a guide, we empirically construct a right triangle with 3″ and 4″ legs and measure to see if the hypothesised 5″ hypotenuse is observed. We find that it is, at least pretty closely!

Nothing could be more elegant, useful, or exciting. We have taken an abstract, uninterpreted theory and shown that it is empirically fruitful. But we do not stop here. We have just begun. We now go back to our metatheory, axioms, definitions and theorems and extend the theory by producing more theorems. We then interpret these theorems to form even more hypotheses. A really fruitful theory will give us many predictions that agree with empirical observations.

Euclidean geometry is obviously useful to physical geographers. But what about human geography? Are theories of this form possible here? Of course they are! Consider central place theory. It results in the theory that a central place settlement pattern is hexagonally packed. By interpreting this theorem, you realise that a hexagonally packed settlement pattern should result in an empirically verifiable R-value (using nearest neighbour analysis) of approximately 2.15 (Clark and Evans, 1954). You can now check to see if the hypothesised prediction agrees with empirical observations. There is no reason to despair if your observations do not agree with the prediction. A negative result means only that you have a lot of scientific work ahead of you, not that fruitful theories and hypotheses cannot be constructed in human geography.

/4/ *I do not agree.* The logical procedures developed by my friends in the Vienna Circle will eventually enable scientists to connect the findings of all the empirical sciences in a single, deductively unified system. The construction of this all-embracing deductive system has already begun and your duty as a scientist is to help nurture it. Your task as a geographer is to identify a portion of this massive deductive enterprise for your special focus. When you choose the topic of your theoretical work, you will want to accept the theorems produced by scientists working on more fundamental parts of the deductive network as basic axioms on which to build your own investigations. As the entire system grows, the principles of physics, chemistry, biology, psychology and sociology will all be linked together logically. Consider the conceptual efficiency this permits! It will be possible to deduce the principles of all the sciences from a master set of fundamental axioms. Since the principles of both human and physical geography will be deducible from this fundamental axiom set, human and physical geography will be logically connected.

I want to emphasise that the principle of unified science was not formulated by physical scientists working in splendid isolation. It was a sociologist in the Vienna Circle, Otto Neurath, who most strongly advocated the cause of unified science. Joergensen (1951, p.76) succinctly summarised Neurath's plan:

> The expression 'unity of science' was introduced into logical empiricism by Neurath. He wanted thereby to mark his opposition to the view that there are different kinds of sciences (and corresponding to them, different kinds of reality or being), such as natural sciences (*Naturwissenschaften*) versus the humanities (*Geisteswissenschaften*), or factual sciences (*Wirklichkeitswissenschaften*) versus normative sciences (*Norwissenschaften*). He also wanted, by the words 'unity of science', to sum up the objective aimed at by logical empiricists, viz., the formation of a science comprising all human knowledge as an epistemologically homogeneous ordered mass of sentences being of the same empiricist nature in principle, from protocol

sentences to the more comprehensive laws for the phenomena of nature and human life.

To reiterate, all scientifically acceptable knowledge can be worked into a deductive network of logically interconnected statements. This possibility is the very foundation of interdisciplinary co-operation and communication. It is literally unjustifiable nonsense to insist that the natural sciences and human sciences deal with different kinds of reality.

/5/ *I agree.* Many of the linguistic traps which give rise to metaphysical assertions have their origin in the use of ambiguous natural languages like English and French. Ambiguity in science is reduced if one translates assertions into the language of symbolic logic. This forces you to be very explicit when forming definitions and sentences. You may argue that something 'gets lost' if all statements in geography are translated into symbolic logic. You are right, of course! You lose imprecision.

Symbolising variables and relationships between variables not only reduces ambiguity, it also enables the process of logical analysis to proceed with greater ease. As the deductive network of unified science grows more extensive and complex, we must continually check the overall logical consistency of the system. This difficult task is much simplified if all scientific statements are expressed in a universal symbolic system. For starters I recommend the notation system developed by Whitehead and Russell (1910-13) in their pioneering work, the *Principia Mathematica*.

/6/ *I do not agree.* Historical time is not relevant to the validity of assertions derived from an uninterpreted theory. In this sense, theories are said to be ahistorical in character. This means that theoretical assertions must hold logically at any time – past, present and future. You can easily see why this is the case. Suppose I offer the following arguments:

(1) A is larger than B, and
(2) B is larger than C

Now, by virtue of (1) and (2) together with the usual meta-theoretical definition of 'and', I offer the following theorem or conclusion;

(3) Therefore, A is larger than C.

It would be absurd to suggest that the logical validity of this
conclusion might change from one historical period to another.
Logically, it had to be true in AD 1066, and it will be true in AD 3045.
The validity of the conclusion does not depend on a specific historical
period. It is ahistorical.

In the same sense, the logical validity of a theory is also aspatial.
If a theorem is logically valid, it must be logically valid in Patagonia as
well as in Iowa. This does not mean that theories are prohibited from
speaking about space and time. It means only that what is asserted about
space and time must be logically valid in all historical periods (including
the future) and all geographical locations.

One consequence of this requirement is that the existence of
historical and regional geography cannot be logically defended.
Geographic theory must be ahistorical and aspatial. A theory of
settlement, for example, must be valid for feudal Europe, colonial New
England, and twenty-first century Africa. A truly scientific approach
to settlement would seek a theory which, when interpreted, would
predict the observable settlement patterns in all past, present and
future regions inhabited by man.

/7/ *I do not agree.* Science involves much more than searching for
empirical generalisations. It is easy (perhaps too easy) to measure
phenomena and then find some statistical shorthand which described
your measurements. The shorthand which summarises your
observations is an empirical generalisation. These generalisations often
permit very precise predictions, but they do not have scientific status
if they were not produced as a result of testing hypotheses which
were deduced from full-fledged scientific theories.

The search for empirical generalisations is often motivated by an
unwarranted metaphysical presupposition: a belief that the world is
inherently orderly and predictable. Many researchers believe that the
task of science is the uncovering of this 'fundamental order'. These
researchers fail to recognise that they may unconsciously impose the
order that they 'discover'. You can find regularity in almost any
pattern if you are clever and persistent. Unfortunately, you can never
know if the order you have found is genuinely fundamental or merely
a product of your way of looking at the world.

It is crucially important to remember that theories do not say
anything about the empirical world. When a theory is fruitfully
interpreted, it will give reasonably accurate empirical predictions. This
does not mean that the world is orderly, only that the theory is

empirically fruitful. Science can never make any statements about the ultimate orderliness of the universe without slipping into metaphysics.

I do not pretend that science can provide explanations in any fundamental or ultimate sense. Yet theoretical science can offer explanations of a kind not attributable to empirical generalisations. Because theorems are always deductively derived, it is always possible logically to account for a theorem in a theory. A theorem accounted for in this way is said to be explained. I can conceive of no other viable meaning for the term 'explanation' without entering the realm of metaphysics. Can you?

/8/ *I do not agree.* Scientific theories are never proven to be true or false. Proofs are found only in the abstract world of logic, not in scientific laboratories. I reserve the term 'proof' for the technical operation of demonstrating that theorems in an uninterpreted theory have been correctly drawn from a set of axioms (and other theorems) according to the rules of logic specified in a given metatheory.

The term proof has no meaning in the empirical world. It is not possible empirically to prove a theory true or false. You may, however, show that a theory is empirically fruitful. If you must choose between two or more alternative theories, you are wise to choose the one which is empirically most fruitful. You must always keep in mind that the empirical success of a theory does not prove that it is true in any ultimate sense. There is always the very real possibility that some future scientist will propose an even more fruitful theory.

You should also recognise that it is largely meaningless to talk about proving a theory false on empirical grounds. Suppose you are attempting to verify the predictions of an hypothesis. Further, suppose that your empirical observations never agree with the predictions. Can you argue that the theory (while it may be logically consistent) is empirically false? Not easily! I can argue that your interpretation or text requires modification. I have you in a box because the number of possible interpretations is nearly infinite. No matter how many unsuccessful interpretations you offer, I can always counter that the next one may be fruitful.

/9/ *I agree.* The claim that hypothesised predictions agree with empirical observations must always be independently verified by many different geographers. This is the famous *principle of intersubjective verification.* It requires that many different researchers (or subjects) must be able to substantiate the claimed agreement between predictions

and observations. This helps to exclude metaphysical assertions from
the accepted body of scientific knowledge. If you do not double-
check the knowledge claims made by others, you have no guarantee
that the knowledge claim is scientifically acceptable. An important
corollary to this principle is the rule that all scientific assertions must
be put in a form which permits empirical investigation.

/10/ *I do not agree.* Perhaps this surprises you? I have insisted that
science must restrict itself to assertions which can be intersubjectively
verified. Does not this mean that I would favour prohibiting all
references to entities which cannot be empirically observed? Not at all!
A prohibition of that kind would make theoretical science largely
impossible. The growth of scientific theory requires careful linkage
between abstract constructs and empirical observations. Attempts to
specify the exact nature of this linkage often resulted in lively debates
within the Vienna Circle. It is still a difficult problem in the philosophy
of science.

Mathematicians have never been troubled by the linkage issue
because they restrict their work to the abstract world of logic.
Mathematics makes no empirical claims. In fact there are mathematical
systems which have no known empirical interpretations. These logical
systems are studied for their own intrinsic interest and for the light
they shed on the nature of logic *per se.*

The opposite extreme describes the activities of systematic
empiricists (Willer and Willer, 1973). Systematic empiricists eschew
theory. Unfortunately, their ranks include many of the so-called 'new'
geographers. Their research efforts are spent trying to find systematic
relationships in empirically collected data. They often employ
sophisticated quantitative techniques, but they are not engaged in true
science. The product of their work is not theory, but a mass of logically
isolated empirical generalisations.

The true scientist embraces both the theoretical and the empirical.
His theory construction is guided by careful attention to precise
definition and the rules of logic. His theories are formalised, axiomatised
and symbolised. He may make reference to unobservable theoretical
entities, but only so long as such references are empirically fruitful.
His empirical investigations are not *ad hoc.* They are carefully planned
attempts to demonstrate the potential fruitfulness of various theories.
In sum, he is more worldly than the mathematician, but his logic is far
more disciplined than that of the systematic empiricist.

Postscript

I am going to leave you with three short questions:

(1) *Were* there ever any 'ideal' positivists in geography?
(2) *Are* there any 'ideal' positivists in geography today?
(3) *Should* there be any 'ideal' positivists in geography?

The questions are brief, but the answers will require considerable digging and reflection on your own. You can begin this task by engaging your professors and associates in conversations designed to test the three hypotheses concerning the hidden nature of positivism in geography. Beyond that, you will want to tap the resources of your library. For a good grounding in positivism proper, you should consult the volumes edited by Neurath (1938-69). Here, in the *International Encyclopedia of Unified Science*, you will find a collection of monographs which explicate the main themes of positivism. Many of these monographs, e.g., Woodger (1939) and Joergensen (1951), are available in separate paperbacks. However, before you dig too deeply into the philosophy stacks, you should realise that philosophy of science is now in a post-positivist era. Good guides to recent developments include: Achinstein and Barker (1969), Radnitzky (1970) and Suppe (1977). Consult Bloor (1975) if you want a list of directed readings in the philosophy of science. If you are responsible for writing a term paper or preparing a seminar presentation on positivism, you will find that *The Philosopher's Index* will lead you to many interesting (but possibly quite technical) sources. This index abstracts the articles appearing in the major philosophy journals. In geography, you will want to begin with a review of two major works which bear the positivist stamp: Harvey (1969) and Amedeo and Golledge (1975). Among the many articles which touch on the role of positivism in geography, you will not want to miss the papers by Walmsley (1974) and King (1976).

Personally, I find that positivism teaches many important lessons. It has taught us the virtue of careful observation. It has also taught us to exercise our minds rigorously and logically, even if we reject the formal structure of theory and the metatheory advocated by positivism. Finally, and perhaps most important, it has taught many of us who once embraced positivism to read more widely in the philosophy of science. Yet no matter what I say or what your professors say and no matter how much you read and discuss, the final decision on the relevance of positivism for your work as a geographer rests with you.

3 PRAGMATISM: GEOGRAPHY AND THE REAL WORLD

John W. Frazier

Introduction

Many factors influence our thoughts, the meanings we assign to objects and events and the actions we take. This is at the heart of philosophical studies, which have interested geographers and other social scientists who seek to understand the nature of reality. In geography we have become especially interested in what particular types of thinking and beliefs influence and guide our individual choice of problems to study, methods to be used and for whom we do research. Many geographers now define and defend particular lines of thought which are borrowed from philosophy.

The purpose of this chapter is to raise some key issues and questions. These will then be addressed within the framework of a particular line of thought. This is not a formal school of thought but rather a set of related ideas extracted from a philosophy termed 'pragmatism'.

Philosophical Questions in Geographic Pursuits

As a social scientist the geographer is interested in explaining and, if possible, predicting the pattern and behaviour of a spatio-temporal phenomenon. The approach used, however, differs considerably. While many geographers do extensive fieldwork, use statistics and utilise cartography in reporting their research, others develop purely theoretical models that approximate certain geographic processes. The point here is that there is a great variety of beliefs which individual geographers hold and a number of fundamental questions must be answered consciously or unconsciously before geographic inquiry begins. The answers dictate how the inquiry will take place and the geographic questions to be answered. For our purposes I will pose three such questions which are pertinent to the philosophical ideas that we will examine:

(1) Is it possible and worthwhile to pursue the development of geographic theories?

(2) How can we best pursue geographic understanding?

(3) How can and should geographic knowledge be used?

In the pages which follow I will outline a general response to these questions which constitutes a position based on pragmatic ideas.

Ideas in a Pragmatic Position

In definitional terms, pragmatism is the 'position in philosophy that defines meaning and knowledge in terms of their function in experience, with reference to adjustment and the resolution of problematic situations' (Beck, 1969, p.515). Pragmatism is largely a Western philosophy and predominantly an American philosophy, although it certainly has had its European proponents. A common position attributed to this philosophy is one of dealing with 'practical' problems. An emphasis of pragmatism has been on the 'practical'. A pragmatist believes that the 'concrete' or the 'particular' situation is important to obtaining scientific knowledge and for understanding the world. While there is certainly some truth to this notion, it results in an incomplete picture of the position. The 'abstract' or 'general' laws and theories are also important and useful as guiding or 'leading principles' in any scientific inquiry. In short, those of the pragmatic persuasion deal with theoretical notions as well as with practical situations. The misunderstanding has come about largely because pragmatists have emphasised 'action' in the way of policy and, therefore, practical solutions to 'practical' problems. This is exactly what many geographers have been supporting as a needed focus for our discipline. For example, one geographer proposed geographic strategies which included 'organisation', 'persuasion' and 'action' to facilitate needed societal transformations (Morrill, 1970). Similarly, another verbalised the need to apply geographic knowledge and expertise to the solving of future problems in conjunction with a necessary transition of society and elucidated the role of the geographer as 'diagnostician', 'prophet' and 'architect' of future society (Zelinsky, 1973). In doing so, he clearly illustrated the need for both theoretical and applied geography in solving future problems. These are but a few examples of the beginning of rethinking research purposes. The product of this inclination was a new desire for dealing with practical questions at a larger scale, beginning the application of geographic heuristics, and a concomitant concern for the nature of geographic education, which, advocates believe, should include practical experiences through internship arrangements with private and public agencies.

Pragmatism as a philosophy was in vogue earlier in this century when it was associated with famous Americans such as Dewey, James and Pierce. This philosophy was popularised by James and his followers. One may wonder why we are now dealing with a philosophy that reached its peak of popularity some years ago. The answer is that good works achieve permanence and continue as useful ideas. This is why today there is a Charles S. Pierce Society among philosophers, who regard this pragmatist 'not as a thinker of bygone days, but as a colleague and co-worker on issues of abiding interest' (Rescher, 1978). Historically, pragmatism grew out of many diverse philosophical viewpoints. As Jarrett and McMurrin (1954, p.253) summarised, it grew out of '(1) the tradition of British Empiricism from John Locke to John Stuart Mill, (2) Kant and his immediate successors, (3) nineteenth-century positivism, (4) evolutionary biology and (5) the new scientific psychology'. Recent interpretations of the pragmatist position are useful in guiding the training of, and in understanding the role of, the applied scientist, including geographers. The following four statements identify the major attributes of the pragmatist's position:

(1) Current reality is a mixture of knowledge and error;
(2) based on our existing fallible systems, we must continue to develop our reasoning on the basis of systematic considerations, experimentation, and re-evaluation – in the world, society and men are in the making with various competing real alternatives as the outcomes;
(3) the scientific method and the hypothetico-deductive model are the best modes of investigation found to date and should be adhered to;
(4) logic should be utilised as a problem-solving device. The problems should be practical and be used for the promotion of human welfare (Aune, 1970).

Each of these points is discussed in some detail below. We believe that such a discussion is essential for an understanding of pragmatism, and would create a suitable backdrop for an assessment of it as a framework for geographic research and teaching.

(1) The Imperfection of Reality. In espousing the imperfections of
reality, the pragmatists were directly reacting to the contention that
the universe or reality is a mechanical universe governed by deterministic
laws which can be observed and calibrated within given subjective errors.
As Scheffler noted, 'the picture is that of nature as an ideal mechanism,
a perfect clock: ideal and perfect' (Scheffler, 1974, p.74). Accordingly,
changes in our knowledge about reality are reflections of changes in
use. The inadequacies or incompleteness of our explanations are not
reflections of the absence of some appropriate principle(s) in nature.
'Blur of measurement', Scheffler noted, 'is a matter of the observer's
limitations rather than the inconsistency of measured magnitudes;
deviations from deterministic formulations are results of our ignorance
and do not indicate that nature itself is anywhere free of the sway of
perfect necessity' (Scheffler, 1974, p.23).

The reaction of the pragmatists to this mechanistic error-free reality
is ably summarised by Charles Sanders Peirce:

> Try to verify any law of nature, and you will find that the more
> precise your observations, the more certain they will be to show
> irregular departures from the law. We are accustomed to ascribe
> these, and I do not say wrongly, to errors of observation; yet we
> cannot usually account for such errors in any antecedently
> probable way (Wiener, 1966, p.170).

Basically, the pragmatists believe that reality is a composite of
knowledge and error.

(2) The Fallibilistic View of Knowledge. Directly related to the
pragmatists' contention of an imperfect world is their view that,
because of the changing nature of reality, and the mind's view of it,
it is impossible to guarantee that an expected outcome would result
from a specific experiment; past successes do not guarantee future
successes. They argue, therefore, that when predictions fail, the
underlying assumptions and hypotheses should be re-evaluated and
modified. If these initial adjustments fail, others should be made.
Although there is no guarantee that an individual would ever find
the critical set of assumptions and hypotheses needed to predict
the future accurately, however, 'the absence of such a guarantee does
not prove that the assumptions we make on a trial basis are completely
unjustified' (Aune, 1970, p.176). This is the fallibilistic view, and
the pragmatists wholly subscribe to it. Aune has summarised the

pragmatists' argument about fallibilism thus:

> The suggestion, therefore, that we are unjustified in making an
> assumption unless we know in advance that it is . . . true is for the
> pragmatist totally unacceptable – as unacceptable as the suggestion
> that a handy tool should be never tried unless one has a guarantee
> that it will prove successful for the task at hand. If we fully realize
> that we are making assumptions on a trial basis and that their
> continued use is justifiable only to the extent that they successfully
> serve their purpose, then our adoption of them can hardly be
> regarded as unreasonable. To assume otherwise is simply to reject
> the experimental spirit of rational investigation (Aune, 1970,
> pp.176-7).

Because of this fallibility the pragmatists underscore the need for
continuous re-evaluation of our assumptions and laws.

Also implicit in fallibilism is the variability of experience. In
philosophical terms, 'truth is attained in a plurality of ways'. The
environment consists of many attributes, the past, as well as the
many varied characteristics of the present. These vary from place
to place and contribute, in an active way, to what we see, hear or
touch. In brief, they affect what and how we experience. These are
the 'contexts' of pragmatism which are believed to result in action.
Again, while we must understand the differences and potential
diverse richness of experience, this is not to deny generalisation.
However, it is these 'particular' or 'practical' contexts which are the
heart of inquiry and against which 'unstiff' theory must be applied.
We must understand that contexts or situations are transient and
changeable. Therefore, for the pragmatist, theory must be a tool,
and policy decisions are judgements based on theory which has
been applied to a situation which is changing, and, therefore,
such decisions must be open to constant re-evaluation and critical
revision; continued enquiry is necessary (Thayer, 1968).

(3) Advocacy of the Deductive-Predictive Approach. As noted above,
the pragmatist uses the deductive-predictive approach. For him,
however, theory serves as guiding or 'leading principles'. More
specifically, a pragmatist utilises theories and constructs hypotheses
which permit him to anticipate, predict and explain events of
experience. Theories are optionally construed as conditional
expressions to understand the objects of experience. In the words

of Thayer,

> To the pragmatist, the conspicuous and significant feature of the
> abstract terms, the definitions, laws, and theories, is the role they
> perform as leading principles. In general, the pragmatist's analysis
> and interpretation of the meaning of theoretical terms consists
> of a general description of the experimental conditions in which
> a certain kind of operation produces a certain set of empirical
> (i.e., practical, conceivable) consequences. The description serves
> as a 'prescription' or 'schema' directing us to the kinds of
> conditions in which a term has its significant use and application . . .
> For the universal statements of laws are meant to function as
> formulations of policy or resolutions. Specifically, they operate
> as decisions to permit certain kinds of inferences from certain
> formulated conditions to others. Fundamental . . . is the part
> played by particular experienced cases − or the possible
> premisses and conclusion of reasoning (Thayer, 1968, pp.376-7).

In this way, the pragmatist deals with the evolution of theory,
hypothesis elaboration, validation and policy decisions over time.

(4) Practical and Purposeful Human Problems. To the pragmatist,
knowledge should be used to solve problems: 'The value of our
intellectual tools must always be measured by their practical success'
(Aune, 1970, p.135). In such situations, they reject the positivistic
viewpoint of 'value-free' research. As Rucker put it,

> . . . it is impossible to study those collective processes without
> perceiving the imminent values. It is impossible, for example, to
> study social organization without perceiving social maladjustments
> or possible economies not realized. It is impossible to study social
> changes without seeing advantageous and disadvantageous
> adjustments. It is impossible also to study the various types of
> social organization without indicating the superiority and
> inferiority of the various types or to formulate a theory of social
> progress without implications of social obligation (Rucker, 1969,
> p.130).

The position of the pragmatist is that enquiry is a suitable procedure
for formulating and testing any type of judgement including a value

judgement. That is, that processes cannot be studied with blinders which block the 'good' or 'bad' of a situation, and predictions include value-propositions which are varifiable.

The above discussion of pragmatism has identified its philosophical tenets. In summary, the pragmatist accepts the relevance of theory as 'leading principles' in the investigation of 'particular' problems. This requires that theory be flexible and that contexts be given careful consideration. It means that deductions from weak preconceptions should be replaced by observation of human activities. The world is open to change and we must continue to test and re-evaluate our positions. This is especially true because of the emphasis in this philosophy on policy formulation and concern for human welfare. Finally, value judgements are an integral part of reality.

There is another way to distinguish between the pragmatic approach and other approaches. It is action-oriented, user-oriented and extends the experimental method to include evaluation and implementation. The research is undertaken for the purpose of solving an immediate problem and results are means to an end for some target population. The researcher gives directives for action and serves as an 'action agent' in the implementation of the results. By the latter I mean the lobbying, persuading or other action which is required in an effort to reach a particular goal. In geography, this approach is distinguishable as planned action rather than thought for planning. Because evaluation and implementation stages are involved, we must deal with the way things should be, not with the way things are. Providing means to ends, which involve human activities and welfare, include values which are an integral part of reality.

The discussion also showed that pragmatism was developed in America. It developed during a period of enormous social and intellectual changes from the Civil War to the beginning of the Second World War. Fisch has adequately summarised these changes:

> The industrialization and urbanization of American society; the exploitation of our natural resources; the spreading and merging of railroads and other systems of transport and communication; the surge toward bigness in industry, business, capital, labor, and education; the management problems of large-scale organization; the drift toward specialization in all occupations; and the rise of an administrative and managerial class (Max Fisch, quoted by Scheffler, 1974, p.3)

Related to these changes were the developments in probability and statistical inference. It was partly in response to these changes that pragmatism evolved as a philosophy around certain basic questions:

(i) How can the new scientific developments in evolution and dynamics be assimilated and used in the understanding of knowledge and knowing?

(ii) 'How are we to connect the life of man with the natural world in which he arises, the knowledge he acquires with the values he espouses, the concepts and abstractions in which his cognition is couched with the realms of willing, feeling, and doing which, no less than cognition, are parts of his life as an organism?' (Scheffler, 1974, p.6).

(iii) What are the new bases of stability in an environment of social, scientific and political change?

(iv) In the light of these developments how can society and the individual be understood?

We have shown, rather briefly, how pragmatism evolved around these four questions.

Pragmatism and Geography

Human interests, desires, prejudices and group values vary across space. Policy based on applied geography, whether it be the modification of an environment, the removal of inequity in housing, or the preservation of a cultural landscape, includes researcher and client values which may vary substantially from the values of other sub-populations, especially those directly involved. The researcher's recommendations can also have long-term impacts. Substantially different policy types depend on which concept the applied geographer gives priority. We must realise how our work is affected when we select an action role, as well as how the research relates to the opportunities of the affected population. One takes a position in an attempt to influence an action, whether it be the location of a facility, environmental change or the spatial allocation of resources. At maximum, moral dilemmas are involved; at minimum, the convenience of sub-populations. Whatever the case, value judgements are involved.

To the pragmatist, the meaning attached to space or movement in space is directly a result of the practical consequences of that space. In the assessment of these consequences, scientific postulates can be made, tested and modified in the light of empirical facts; the approach is purely inductive. For the pragmatist geographer, spatial laws are valid, and they provide the framework for hypotheses formulation and data collection. Furthermore, hypotheses about spatial structure can be formulated, tested and modified in the light of empirical evidence.

The adoption of some aspects of the scientific method by pragmatists means that their method of analysis is a modification of the more rigid positivistic views of research. Their rejection of complete determinism underscores their strong belief in the resolution of geographic problems by constant adjustments and modifications of the hypothesis, in the light of empirical data.

As a strong empirical based philosophy, pragmatism is also a humanistic philosophy. Dooley's (1974) recent book on William James, one of the 'founders' of this philosophy, stressed this attribute of pragmatism. To James and the other pragmatists, the human element was important indeed. As James put it, 'it is impossible to strip the human element out from even our most abstract theorizing' (James, 1904, pp.175-6). This view is similar to those expressed by Paul Vidal de la Blache and the French School of Geography. The aim of pragmatism is to emphasise the human element; 'Our thoughts determine our acts, and our acts determine the previous nature of the world' (James, 1932, p.318). Here man is central. In humanistic geography, man and science are reconciled. As Ley and Samuels put it, 'a principal aim of modern humanism in geography is the reconciliation of social science and man, to accommodate understanding and wisdom, objectivity and subjectivity, and materialism and idealism' (Ley and Samuels, 1978, p.9).

From the above discussion some of the elements of a pragmatic geography can be identified:

(i) Geographic space is a composite of knowledge and error.

(ii) Geographic space is changeable as our knowledge of it changes and the scale of measurement becomes more refined.

(iii) Geographic space is a manifestation of the 'human element' through time.

(iv) Geographic space is structured and restructured as a result of solutions to practical human problems.

(v) Spatial reality is a composite of human experience. As Dooley noted, 'the inability of human experiences to be objective (i.e., neutrally disinterested) does not preclude our knowing the real, for man and his activities are part and parcel of reality' (Dooley, 1974, p.173).

(vi) Spatial laws are useful for hypothesis formulation, but the hypothesis may be modified in the light of our knowledge.

(vii) Geographic studies are concerned with the practical problems of man in space.

(viii) Geographic problems can be studied using the scientific inductive method.

Although the pragmatic theme may be implicit in many geographic studies (see Morrill, 1970; Zelinsky, 1973), only a few have been guided by this philosophical viewpoint. An example in point is a recent paper by Frazier and Budin (1979) on the application of innovative behaviour theory to the assessment of a housing rehabilitation programme in Johnson City, New York. This project is discussed in some detail in order to illustrate some of the points raised earlier in this chapter.

(a) The Problem. The blighting of older single-family homes is a common problem facing American cities. Local communities through federal financial assistance have developed programmes to assist lower-income households rehabilitate their homes in an effort to eradicate the blight condition and create pleasant residential areas in the city. Johnson City, New York, has a Community Development Department (CDD) which administers such a programme. The CDD received, in 1978, a quarter of a million dollars from the Housing and Urban Development Office in Washington, DC, to assist local residents in housing rehabilitation efforts.

The Community Development Department set its goals as: (1) assisting low-income families, established by a formula which considered total household income and family size; and (2) eradicating the visible external blight. When the first programme year was nearing its end, the department director and staff became puzzled and discouraged because few lower-income households had applied for assistance.

Spatially, the applicants were concentrated in a few neighbourhoods, which tended to be in the best physical condition. At this point, looking forward to the next programme, the CDD invited Frazier and Budin to participate in analysing the problem and making recommendations to change the pattern.

(b) Spatial Laws and Theories Used. In dealing with this particular case, they used the principles of diffusion and innovative behaviour theory. This theoretical approach links propagator and adopter of an idea, event, or programme into one framework for the purpose of theoretical clarity. Of particular relevance to them, was the principle that the spread of an innovation from an agency to its local market area is largely dependent on the actions of the agency which 'mould' the patterns of adoptions, while households 'detail' the broader pattern through individual actions (Brown, 1973). Generally, according to diffusion theory, the acceptance of any new programme or product depends on price, cost and information.

(c) The Possible Cause of Low Low-Income Participation. An examination of the methods used by the Agency for the dissemination of the information about the programme included a complex brochure, local media news broadcasts and 'town meetings', which were advertised by the brochure and media, as well as 'posters' placed in public places. These were not media for reaching the poor. From such an examination they concluded that the approach used for the spread of the information needed changing.

(d) Some Working Hypotheses. They proposed the following hypotheses:

(1) there is a significant difference between lower income and higher income groups' awareness of the housing rehabilitation programme's existence; and

(2) a personal agent methodology for 'delivery' of the programme can and should result in programme subsidies to the poorer groups.

Implicit in the first hypothesis was that information was not reaching a significant number of poorer households who needed assistance. The second asserts that the poorer households would behave as theorised; they would not only receive information but would and should be made aware of how this particular programme was relevant to their needs. An expected result of the proposed methodology, then, was the elimination of the bias of programme applications in select neighbourhoods.

(e) The Results. In consultation with the CDD, these hypotheses were tested. The test projects confirmed the hypotheses and a larger scale operation with the new policy followed. While some serious problems still remain for the housing rehabilitation programme, the result was a substantial increase in the numbers of lower-income households that applied for participation in the programme.

In the above example, the authors followed the basic tenets of the pragmatist philosophy:

(i)   theoretical propositions were used in guiding explanations and recommendations,
(ii) the scientific method was utilised in solving the problem,
(iii) the particular context of the poorer households was given careful consideration,
(iv) the CDD's approach was judged as being inadequate and wrong for poorer households and the authors strongly advocated a particular approach as necessary and good for the lower-income households.

All individual philosophies are extremely complex. A short chapter, such as this one, cannot possibly deal successfully with all aspects of a given philosophy. However, the chapter has dealt with a few fundamental ideas of pragmatism which are offered not as a formal school of thought, but as a viable alternative to the existing philosophical viewpoints for the study and analyses of spatial systems. Pragmatism is appealing because it combines the scientific method with humanistic geography.

4 FUNCTIONALISM

Milton E. Harvey

Introduction

Of all the philosophical viewpoints that have both directly and indirectly impacted upon geography, functionalism is one of the few that has largely evolved as a result of persistent theoretical, philosophical and empirical debates among social scientists – especially sociologists, anthropologists and, more recently, political scientists. In spite of its 'social scienceness', functionalism as a research and pedagogical vehicle has not significantly influenced geography. Implicitly, however, many geographic studies have followed a functionalist mode. It is this author's contention that many geographic problems can be understood and studied in the framework of functionalism. Furthermore, functionalism may provide a philosophical and methodological umbrella for many diverse topics in our discipline. This chapter begins with a brief historical exposé of the changing meaning of functionalism. In subsequent sections, the advantages and disadvantages of functionalism are discussed and the links between aspects of functionalism and geography are explored.

What is Functionalism?

Very simply, functionalism is concerned with functions, and the analysis of functions of certain customs, acts and artifacts to the society. As Gouldner noted, functionalism is nothing if it is not the analysis of social patterns as parts of larger systems of behaviour and belief (Gouldner, 1959, p.241). Recently, Eisenstadt and Curelaru noted that functionalism:

> stresses the systemic properties of groups, institutions, and macro-societal orders, their internal organizational and structural characteristics or dynamics, and possibly, their interrelationships. It has tended to define social units as systems or organizations with specific structures and needs (Eisenstadt and Curelaru, 1976, pp.86-7).

In political science, the importance of group properties and super-international organisations designed to solve specific human problems is at the heart of functionalism. Functionalism, in this context, focuses on the structure and functions of international organisations within a framework of working peace that, in the words of Myrdal, would build the Welfare World.

The above discussion was an initial attempt to give the student some idea of what functionalism entails. Like any other conceptual and philosophical viewpoint in the social sciences, definitions of functionalism have varied over time and across disciplines in response to both the needs of that discipline and criticisms about the limitations of time-specific prevailing viewpoints. In general, the major developments in the definition of functionalism are associated with the works of Emile Durkheim, Bronislaw Malinowski, Alfred Radcliffe-Brown, Talcott Parsons and Robert Merton in sociology and anthropology. The contributions of each to functional theory are briefly discussed below.

Emile Durkheim and Functionalist Ideas

Functionalist ideas are usually traced to Durkheim who developed an organicist's view of functionalism. He regarded society as an entity; a system, in space and time, which cannot be reduced to its constituent parts. In a broader context, he 'viewed system parts as fulfilling basic functions, needs or requisites of that whole' (Turner, 1974, p.18). He contended that the cultural practices of a society are a function of the social organisation of that society. As noted in a later translation of his work, '. . . all moral systems practiced by peoples are a function of the social organisation of these peoples, are bound to their social structures and vary with them' (Durkheim, 1938, p.67). His major contributions to functionalism were the systemic view of society, the functional needs of society, the roles performed by the constituent parts of the society, the implicitness of societal equilibrium, and the importance of causal relationships.

Malinowski's Functionalism; Teleological Reductionism

The influence of Durkheim on Malinowski is evident from Malinowski's earlier definition of functional analysis in anthropology as:

> explanation of anthropological facts at all levels of development by their function, by the part which they play within the integral

system of culture, by the manner in which they are related to each other within the system and by the manner in which this system is related to the physical surrounding (Malinowski, 1926, p.132).

From the above quotation, certain attributes of Malinowskian functionalism are evident. First, the emphasis is on the functional role of the constituent parts of a society. Like Durkheim, he implicitly assumes that every social activity or artifact had a functional need in that society. Second, the emphasis is on the functional interrelatedness which implies sub-system linkages, sub-system feedbacks and system equilibrium. Third, he specifically underscores the need to study and understand a culture in its environment. He insisted that every cultural institution must be evaluated and understood within the framework of the cultural milieu in which it developed. As he noted, 'human beings live by norms, customs, traditions, and rules, which are the result of an interaction between organic processes and man's manipulation and re-setting of his environment' (Malinowski, 1944, p.68). This man-environment view parallels the *genre de vie* of Vidal de la Blache and the French School of Geography.

We cannot present a discussion of Malinowski's functionalism without some comment on why it is regarded as one of teleological reductionism. Malinowski, like Durkheim, believed that every social entity or component performed a needed unique role in the maintenance of the society. This, of course, is illegitimate teleology. The reductionistic tendency of Malinowski is manifest in his analytical scheme of society. Every society is primarily concerned with the fulfilment of human primary imperatives of food, shelter and reproduction. To meet these basic needs, organisations and communities (with concrete iconographies) are necessary. In turn, these organisations create additional needs which are fulfilled by the creation of more complex social and political organisations. From such a cascade of process-response interlinkages, Turner argues,

. . . it is possible to visualize several types of requisites shaping culture: (1) those that are biologically based; (2) acquired psychological needs; and (3) derivative needs that are necessary to maintain the culture and patterns of social organization which originally met basic biological and acquired psychological needs . . . By visualizing culture as meeting several 'layers' of such requisites, Malinowski could employ reductionist argument to explain the existence and persistence of any structure in society (Turner,

1974, p.24).

In Malinowski's functionalism, fieldwork was the only way to collect data on behaviour (Malinowski, 1944, p.71).

Radcliffe-Brown's Functionalism or Structuralism

Starting from a discussion of function and what he termed the 'necessary conditions of existence', or more simply 'needs' in a purely physiological context, Radcliffe-Brown developed his functional-need relationship in social life. He asserted that

> the function of any recurrent activity, such as the punishment of crime, or a funeral ceremony, is the part it plays in the social life as a whole and therefore the contribution it makes to the maintenance of the structural continuity (Radcliffe-Brown, 1952, p.180).

From the above brief comment on Radcliffe-Brown's functionalism, certain themes emerge. First, unlike Malinowski, he focused only on recurrent activities such as marriage, kinship and religion – 'science (as distinguished from history or biography) is not concerned with the particular, the unique, but only with the general, with kinds, with events that recur' (Radcliffe-Brown, 1952, p.192). The second attribute of Radcliffe-Brown's functionalism is that of linkage systems; at both the inter-personal and the activity level. The third attribute is that recurrent activities do perform essential functions for the whole society. It is from the study of such functions that hypotheses may be developed and tested within a nomothetic framework. As he put it:

> we formulate a hypothesis as to the nature and function of ritual or of myth. This requires to be tested, and may ultimately be proved, by a sufficient series of studies of cultures of different type, in each of which the whole system of ritual or of myth has to be considered in its relation to the culture as a whole (Radcliffe-Brown, 1977, p.45).

This is what he called the functional hypothesis. Fourth, unlike Malinowski, Radcliffe-Brown's functionalism attempted to deal with the problem of disorder, dysnomia, and change in a society. He argued that unlike organisms that die from extreme situations of disorder, the society may not die, 'it can change its structural type, or can be absorbed as an integral part of a larger society' (Radcliffe-Brown, 1952, p.182). Finally, Radcliffe-Brown believed that social structure encompasses both social morphology and interaction. From social structure Radcliffe-Brown logically moves to the concept of structural systems: 'There is such a thing as social structure', he asserted, 'and the theory of social evolution depends on this concept' (Radcliffe-Brown, 1977, p.51). This is in contrast to Malinowski's basic organismic viewpoint. Radcliffe-Brown strongly stressed this difference: 'This theory of society in terms of structure and process, interconnected by function, has nothing in common with the theory of culture as derived from individual biological needs' (Radcliffe-Brown, 1977, p.32). Because of his emphasis on social structure, he preferred structuralism to functionalism.

Talcott Parsons – Functional Imperativism

To Parsons, the systems model implicit and relatively underdeveloped under the aegis of the early functionalists, was made explicit. To him, functionalism is concerned with the conditions for pattern maintenance of an organism and that organism's interchanges with the environmental system. As he put it:

> For any system of reference, functional problems are those concerning the conditions of the maintenance and/or development of the interchanges with environing systems, both inputs from them and outputs to them . . . Function is the only basis on which a theoretically systematic ordering of the structure of living systems is possible (Parsons, 1977, p.180).

To Parsons, functionalism implied certain basic themes and assumptions: (1) that the social system operates in the socio-cultural milieu (or environment); (2) that the social system, like organisms, need to survive, and that certain functional prerequisites are essential for this survival; (3) that the intra-social system organisation of constituent activities is a result of the structured responses of the social system to the functional prerequisites.

In formulating his systems model of functionalism, Parsons strongly advocated the use of analog theory. In a recent paper written as part of a *Festschrift* volume in honour of Robert Merton, he asserted that 'the existence of fruitful analogies between the phenomena of organic life and those of human personalities, societies, and cultures rest essentially on the common features and continuities of different types of living systems' (Parsons, 1977, p.110). To Parsons the two major areas of analogies between human society and biology are the fundamental concepts of *adaptation* and *integration.* These are the result of the interaction of the environmental and the behavioural systems. The environment encompasses more than just the physical sub-system: it also includes the personality sub-system, the cultural sub-system, and the organism sub-system. The impact of the physical sub-system is manifested in the scarcity of resources, that of the cultural sub-system in cultural or pattern-maintenance and goal attainment, whereas the personality sub-system, or more specifically the personalities of the members of the society, is closely related to the functional prerequisites of integration and latency.

The four functional prerequisites of adaptation, goal attainment, integration and latency directly impact upon the behaviour of the members of the social system. As Jessop noted, the 'social order depends on the continuing fulfilment of the four functional problems and also on the maintenance of balanced relations between the social system, and the other systems of action and the physical environment' (Jessop, 1972, p.18).

From the brief analysis of the major aspects of Parsons' functionalism, it is evident that to Parsons the analysis of complex systems started with the isolation of the basic components that constitute them. This isolation should be followed by an investigation of the inter-linkages of these components. The complications caused by these inter-linkages should be studied by functional analysis; hence Parsons' functionalism has been labelled functional imperativism. He stressed the interdependency of the independent sub-systems, and how they function toward the goal of maintaining the system.

Robert Merton – Functional Structuralism

For Merton, the major orientation of functionalism is 'expressed in the practice of interpreting data by establishing their consequences for larger structures in which they are implicated (Merton, 1957, pp.46-7). He argued that because of the paucity of adequate empirical groundwork, sociology was not ready for the type of systems models suggested by Merton. Functional analysis, he argued, should concentrate on some unit of human behaviour so as to explain why it persists in a specific society.

As a backdrop to the development of his paradigm, and protocol for the construction of functional theories of the middle range, he modified the basic tenets of what is usually referred to as the trinity of functional postulates: functional unity, the universality of functionalism and the indispensability postulate. Briefly, the postulate of functional unity assumes that recurrent activities are indeed functional for the whole social system. Merton argued that 'this unity of the total society cannot be posited usefully in advance of observation. It is a question of fact, and not a matter of opinion' (Merton, 1957, p.30). The second postulate, that of universal functionalism, asserts that all recurrent cultural items or acts have positive functions in that society. In place of this, he suggested the 'provisional assumption that persisting cultural forms have a net balance of functional consequences either for the society considered as a unit or for sub-groups sufficiently powerful to retain these forms in fact' (Merton, 1957, p.32). The third postulate – indispensability – alleges that certain functions and existing social institutions are indispensable. In modifying these two concepts, Merton stressed that 'just as the same item may have multiple functions, so may the same function be diversely fulfilled by alternative items' (Merton, 1957, p.34).

It was in the context of these criticisms that Merton proposed 'a paradigm of functional analysis in sociology'; the purpose being 'to supply a provisional codified guide for adequate and fruitful functional analyses' and 'to lead directly to the postulates and . . . assumptions underlying functional analysis' (Merton, 1957, p.55). Under this paradigm, the question of needs and requisites fulfilled by an item or custom can only be investigated after (a) the item has been identified as being recurrent; (b) the motivation of the individuals in the social system toward the goals have been understood; and (c) an understanding of the functional and/or dysfunctional consequences of the item has been determined in terms of both manifest and latent functions. In this framework, systems analysis is implicit; 'either by treating the structural context to which the unit is linked as a system, and/or by analyzing the unit itself as a subsystem composed of interdependent parts' (Gouldner, 1959, pp.243-4). Because of this emphasis on the structural context, Merton's functionalism has been labelled 'functional structuralism'.

Other Functionalist Viewpoints

As shown above, there is divergence of views on functionalism; a reflection of differences in interpretation of words, and differences in

discipline. Let me elaborate. The word 'function', which is the key
ingredient to functionalism, has been interpreted in five major ways:
(a) it refers to a public gathering for a specific ceremonial purpose;
(b) in Weberian terms, and in geography, it is synonymous with
occupation; (c) in political science, it refers to the duties associated
with a job that involves the exercise of authority; (d) in a mathematical
sense, it refers to the relationship between a variable and another; and
(e) as in functional analysis in sociology and biology, it refers to the
processes which contribute to the maintenance of the organism.

The diversity of definitions has resulted in diversity of meanings
of functionalism within a discipline and in the various social sciences.
The diversity of views in sociology has already been discussed. In
anthropology, to give another example, scholars have never really
incorporated the valuable works of the sociologists like Parsons and
Merton. Most of the 'polemics' have focused on the limitations of
the so-called 'Malinowskian Dilemma' in cross-cultural inquiry. The
place-specific nature of Malinowski's approach is not useful for
cross-cultural studies because it leads to the comparison of
incomparables. Modifications of Malinowski's approach that would
make it suitable for such cross-cultural studies have been suggested
by Goldschmidt (1966)..

In political science, the origin of functionalism is generally
associated with the work of David Mitrany in the 1930s. This form
of functionalism is based on the principle that:

> . . . man can be weaned away from his loyalty to the nation
> state by the experience of fruitful international cooperation; that
> international organization arranged according to the requirements
> of the task could increase welfare rewards to individuals beyond
> the level obtainable within states; that the rewards would be
> greater if the organization worked, where necessary, across
> national frontiers, which very frequently cut into the organization's
> ideal working area. Individuals and groups could begin to
> learn the benefits of cooperation and would be increasingly
> involved in an international cooperative ethos, creating inter-
> dependencies, pushing for further integration, undermining the
> most important bases of the nation state (Taylor and Groom,
> 1975, p.x).

Basically, this form of functionalism focuses on the interlinkages
between states and has many common attributes to Deutsch's

transactionalism which was introduced to geographers by Soja (1968). It believes that international organisations which emerge to 'satisfy felt needs' (Taylor and Groom, 1975, p.1) would lead to greater benefits for individuals. This 'satisfaction of felt needs on a non-national basis' would eventually undermine national allegiances; the central axiom being that 'form should follow function' (Taylor and Groom, 1975, p.1). Central to this functionalism, as to that in sociology, is the interaction of systems or sub-systems, the existence of needs to be fulfilled, and the importance of actors to perform 'requisite' functions that would result in the creation of a 'better', larger form of spatial organisms or groups. To the political scientist these concepts come into play not because of a recurrent activity but because 'a working peace system exists and functionalism seeks to remove impediments to its further growth' (Taylor and Groom, 1975, p.2). From this idea of larger more efficient (spatial) structures created by adjustments in the size of the units, this form of functionalism easily assumes that:

(i) it is proper for groups larger than the family to provide essential welfare needs of individuals;

(ii) since, in the evolution of the nation-state, the family was superseded by the clan, and the clan by the tribe, the tribe by the state, then the state must ultimately give way to international organisations which better provide the welfare needs.

These tenets have been applied to diverse topics such as international political integration, regional integration and international economic relations (See Sewell; Taylor and Groom; and Haas).

In mass communications the use of functionalism has focused on the satisfaction of needs. Specifically, the satisfaction of needs and need gratification are used to explain media behaviour and newspaper readership. To quote Mendelsohn, 'predisposing needs often are referred to as explanations for a variety of observed media behaviour. Newspaper readership may be explained, for instance, in terms of needs for linking community and society. Assiduous viewing of televised football games ventilates repressed hostility' (1974, p.387). Needs, wants and expectations affect the uses to which individuals put the media.

In sociology, so widespread was the functionalist approach that many scholars have regarded it as a paradigm (Friedericks, 1970, pp.18). To Kingsley Davis, however, functionalism was basically concerned with topics that also concern all sociologists. Thus, sociology and functionalism were synonymous. As he put it, 'every science describes and explains phenomena from the standpoint of a system in nature. In the case of sociology, what is distinctive is the subject, not the method' (Davis, 1967, p.381).

What emerges from the above brief discussion is the belief by functionalists that needs exist and their fulfilment is essential for the organism. The fulfilment requires the identification of goals and approaches for solving certain requisites. Specifically, the basic principles of functionalism are (synthesised in part from lists presented by Goode, 1973, pp.72-3, and Van den Berghe, 1967, pp.294-5):

(i) Societies should be examined holistically in an interrelated systems framework, within largely spatially conformable units.

(ii) Causation is reciprocal and, in many instances, multiple.

(iii) Social systems are generally in a state of dynamic equilibrium. The influence of external forces do not significantly impact upon this stability.

(iv) Because of (iii), change is generally gradual and reflects the influence of adjustment to external inputs, growth through more functional and structural specialisation, and innovations from within the system.

(v) The functionalist is less interested in the history of a society, but more concerned with social interaction. They are more likely to focus on the role and meaning of certain societal functions or subjects such as rituals, dances and symbols.

(vi) Functionalists attempt to find the interrelationships between the components of a social structure.

(vii) In fieldwork the functionalist emphasises:

(a) An account of the environment and the major actors with a special emphasis on their interrelationship within the society.

(b) A detailed documentation of the recurrent function, how it is performed, and a consideration of what is ignored.

(c) An account of the motive and emotional meanings which the actors attach to the ceremony, act or artifact.

(d) A description of any regularities that may be observed by the researcher. As Merton noted, 'the inclusion of these "unwitting" regularities in the descriptive protocol directs the investigator almost at once to analysis of the pattern in terms of what we have called latent functions' (Merton, 1957, p.60).

Functionalist Methodology

When functionalist methodology is articulated, there is a tendency to advocate a biological organismic model within the framework of systems theory. Developed as a scientific model of social systems, functionalism was impacted by the computer age – reflected in the use of sophisticated analytical techniques that were previously the domain of economics. One technique that has been adopted is game theory. Because the concepts of game theory bear similarities to those of functionalism, it 'was double attractive, for it enabled sociologists and cyberneticians to link two or more sub-systems into larger contextural systems in a manner analogous to the way in which human agents adjusted to their responses to the action – or presumed intended action – of others' (Friedericks, 1970, pp.16-17).

The implicitness and, in some cases, the explicitness of systems theory in functionalism has made systems methodology ideal for functionalism. At one level, this has entailed conceptualisations designed to capture the pattern of interlinkages between the major sub-systems. Most of Parsons' work, for example his articulation of the dyadic linkages between his four prerequisites, was at this conceptual level. At another level, it entails the formulation of a system of linear equations that capture the interactions between the major sub-systems. This approach is, in another context, advocated by Wilbanks and Symanski (1968).

It is generally agreed that functionalism is a type of theoretical explanation (Nagel, 1961). The precise nature of this explanation, however, is one of debate. Some scholars argue that it is 'indeed a valid method of explanation, but not a distinct method since all science, by studying the relation of parts to the whole, follows the same procedure as functionalism' (Isajiw, 1968, p.6). Some others, including Hempel (1959), contend that functionalism is a distinct method of explanation. However, they argue that it is a weak and inadequate method of explanation. Some of Hempel's criticisms are discussed later in this chapter. Another view is expressed by scholars such as Homans (1964) who argue that functionalism is a research method but not a method of explanation.

In an attempt to formally analyse functional methodology, Isajiw summarised functional methodology thus:

b has the function of C for a system d.

where b is a specific cultural act, entity or custom in the self-persisting, boundary maintaining system, d; and C is the needs that b is supposed to satisfy (Isajiw, 1968, p.29). The basic question that is inherent in the above is whether it implies that

$$\text{or} \quad \begin{array}{l} b \text{ is a cause of } Cd \\ Cd \text{ is a cause of } b. \end{array}$$

In exploring these possibilities, Isajiw used the principles of productive causality. Briefly, this principle asserts that if a particular entity, say b, is observed, then another event, situation or act, say d, is also observed. This implies that b is the cause and d is the effect, or b produces d; if b then d. Characteristic of this type of causality are the concepts of conditionalness (that all the conditions essential for the occurrence of b are satisfied), adequacy (that the occurrence of b is indeed adequate for the occurrence of d), invariability (that if there is ab, ad would invariably occur), uniqueness of bond (that the occurrence of d follows 'in a unique and unambiguous way' from the existence of b') (Isajiw, 1968, p.32), and continuity of action (that the existence of ab gives rise to d and only d). Of these, the last two are absent in functional methodology, and its causality implies 'b is a cause of Cd'; the parts of the system are related to the needs of the system. Because of this, Isajiw concluded that 'inasmuch as it provides an explanation, however, it assumes that the explananses are adequate to account for the explanandum' (Isajiw, 1968, pp.33-4). As Isajiw later concluded, 'of all the characteristics of productive causality, functionalism involves only the sufficiency aspect, but lacks the necessity aspect' (Isajiw, 1968, p.40).

Basic Criticisms of Functionalism

Functionalism has been criticised on both conceptual and logical grounds. Conceptual criticisms include conservatism of viewpoint, preoccupation with states of equilibrium (a *status quo*) and the assumption of total social (spatial or global) integration. On the logical grounds, the basic criticism is that of teleology.

On conceptual grounds, critics of functionalism contend that the preoccupation with society as a system does not allow functionalists to study or focus on such contemporary problems as poverty, the rapid Westernisation of a traditional society, war, diseases and racism.

Gouldner has specifically accused Parsons of overemphasising the *status quo*, and therefore focusing on 'ways in which it (the status quo) was open to change rather than the manner in which its own characteristics were inducing the disorder and resisting adaptation to it' (Gouldner, 1970, p.147). Because of this invariance to change it will not appeal to the present generation of sociologists; 'it is not likely that the devotees of Psychedelic Culture will find Parsonsianism congenial; indeed, the mind boggles at the thought of a Parsonian hippie' (Gouldner, 1970, p.160).

The equilibrium tendency implicit in functionalism has also been criticised. In functionalism equilibrium is the result of the integration of the system in terms of a centralised value system. Because of this tendency to explain social systems in terms of no change and the importance of roles and controls to the maintenance of the *status quo*, functionalism has been accused of advocating social control rather than social change. Such a view, Kolb contends, is contrary to the assumptions of Judaic-Christian doctrines on which Western societies were built (Kolb, 1961). In response to the criticisms, Parsons noted that:

it [functionalism] has nothing to do with political conservatism or a defense of the status quo. It has nothing essentially to do with judgements about the specific balances between elements of integration in social systems and elements of conflict and/or disorganization. The concept 'dysfunction' is, of course, just as legitimate and important as that of function in the positive sense . . . A related polemical orientation is the claim frequently put forward that 'functionalists' are incapable of accounting for social change: that is, their type of theory has a built in 'static' bias. This also is entirely untrue. If we have any claim to competence as social scientists, we must be fully aware that there are problems of positive integration and malintegration. The student's orientation to these problems is not a matter of the type of general theory he subscribes to but of his more empirical interests and his empirical judgements (Parsons, 1977, pp.108-9).

Another criticism is that many functional explanations are structural in nature. That is, the explanation of an observed pattern does not make reference to the underlying motives or processes; it is largely in terms of sub-system interrelationships. The preference by Merton and Radcliffe-Brown for structuralism over functionalism clearly underscores their emphasis on structure rather than processes.

Another conceptually related criticism of functionalism is the absence of definitional clarity. One of the best examples is the use and definition of 'needs'. What constitutes a need? Do needs have the same significance to different actors? What would happen if needs are not fulfilled? Will the society die? These difficulties have given rise to a diversity of meanings of need. As Jarvie noted:

> Needs could, for example, be relativized to survival, thus unless the needs or requisites of the society are met, it will not survive. But then, we must have some specification of what it is for such a system to live, what its normal or equilibrium state of living is like, how it is to be recognized and tested for. This has hardly been done, and the result is quite diverse subjective interpretations of 'need' statements (Jarvie, 1973, p.27).

The multiple definition of 'functions' discussed earlier adds to the ambivalence of meaning in functionalism.

On logical and methodological grounds, one of the major criticisms against functionalism is that of teleological explanation. Briefly, teleological explanation explains a given situation 'not by reference to causes which "bring about" the event in question, but by reference to ends which determine its course' (Hempel, 1959, p.277). That is, 'the end is the "cause" of the steps toward it' (Goode, 1973, p.74). Some examples would make the point clear. Goode gives this rather simple example: 'vultures were created by nature in order to get rid of corpses' (Goode, 1973, p.74). One of the most commonly quoted examples of teleology is Malinowski's account of the function of magic: 'magic fulfills an indispensable function within culture. It satisfies a definite need which cannot be satisfied by any other factors of primitive civilization' (Malinowski, 1954, p.90). In these examples, the implication is that vultures and magic are indispensable for the specific functions they perform. As Hempel rightly noted, however, there are alternatives that would equally fulfil these functions. At a broader level, this implies that the assumption of functional indispensability, associated with recurrent socio-cultural constituents, may be more appropriately replaced by an assumption of functional substitutability. The function of a vulture can be efficiently performed by others, such as foxes, lions and man. Although this principle of functional substitutability was insisted upon by Merton, however, the substitutes must be from either the cultural milieu or the environment, otherwise the adoption of the substitute would effect changes in the system that would disrupt the *status quo* and change the system. Substitutes with such effects are *not* regarded as functional equivalents.

An example would help elucidate the last two sentences. The substitution of an 'imported' currency system for a barter system, in a technologically backward society, has the far-reaching implications of technological change, changes in clothing, and even changes in the traditional social order. Such a substitution would not be regarded as equivalent, because 'as a result of adopting the modified pattern the group had changed as strongly in regard to some of its basic characteristics . . . that it was not the original kind of primitive group any more' (Hempel, 1959, p.285).

Besides the problem of functional substitutability, teleological arguments have also been criticised on their assumption of self-regulation. This 'hypothesis of self-regulation can only be used as a basis for explanation and/or prediction if it is a reasonably definite statement that permits of objective empirical test' (Hempel, 1959, p.291). As noted above, most functionalist studies do not clearly define the type of system to which the self-regulation applies, nor are the scopes of frequently used terms adequately specified.

The teleological criticisms of functionalism have largely been discussed by philosophers of science, many of whom, especially Hempel, contend that if functionalism is a causal explanatory model, then a positivistic deductive nomological approach to scientific explanation is the natural method to adopt (See Harvey, 1969 for explanation). To Brown (1968, p.110), functional statements may become explanations if they were limited to self-regulating systems with feedback mechanisms. As he noted, 'functional relations hold only between traits within a specified system of a certain type – a self-persisting one . . . Functional relations, then, are certain causal ones which operate within self-persisting systems'. Although such systems exist in geography, Harvey argues that because geographers accept the use of functional explanations, 'we require to specify the precise nature of the self-regulatory system and to show, by way of empirical test, that this specification is a reasonable one' (Harvey, 1969, pp.436-7).

Sociologists have generally dismissed these criticisms as a result of an inadequate understanding of functionalism. As Goode reiterated, 'some critics are philosophers of science, like Carl G. Hempel, Ernest Nagel, and Dorothy Emmet, who have simply not read enough in the field to understand the issues' (Goode, 1973, p.65). In spite of this attempt to ameliorate their impact, many sociologists feel that it is no longer a viable viewpoint (See Gouldner, 1970). Others believe that since the tenets of functionalism are basically equivalent to those of sociology, it is tautological to talk about functional sociology. A third viewpoint, recently proposed by Merton (1976), is that functionalism is one of a series of approaches in sociology.

The criticisms aside, functionalism has advocated an approach which is presently being used as a research format (sometimes in a modified form) in many social sciences. As Harvey rightly observed, 'the logical deficiencies of functional explanation . . . [do] not in any way vitiate the use of functional statements in explanation. Taken in conjunction with certain other specific conditions, function statements may indeed be used in the course of offering an explanation' (Harvey, 1969).

Functionalism and Geographic Investigation

Functionalism has entered into geography in many direct and indirect ways. Indirectly it has entailed the use of functionalist concepts; directly it involves attempts at using some form of functionalism.

Indirect Functionalism

For purposes of simplification, we have isolated six interrelated concepts that are an integral part of functionalism:

(a) Functions
(b) Functional Substitutability
(c) Goals
(d) Pattern Maintenance
       Self-Regulation/Status quo
(e) Adaptation
(f) Integration

Functions. Each of the five definitions of functions discussed earlier has been widely used by geographers. In the studies of the functional classification of cities, the implicit meaning of functions has been occupational. The number of people employed in the various occupational categories have constituted one of the major data sources for such studies. The importance of secondary occupational groups to the continued growth of the city are implicitly functional prerequisites for city growth. Without such activities, the city either declines, becomes moribund or changes its emphasis from secondary to quaternary occupations. However, since many cities in the Third World grew without any appreciable growth in secondary occupation, we have to conclude that such functional requisites are necessary but not sufficient conditions for city growth.

When defined as jobs of a body (say a mayor) or organisation (say corporations), function has entered into many diverse research areas of our discipline. For example, political geographers believe that the functions of a state are to protect itself from external aggression, and ensure internal law and order according to the legal codes of that state. The third meaning of function, defined as the relationship between a dependent variable and one or more independent variables, is very common in geography. The fourth meaning of function is that of role. This shade of meaning is common in studies of systems of cities and associated activities. Cities have roles designed to enhance both intra-regional and inter-regional interactions. Stohr refers to these two roles as intra-regional functions and extra-regional urban functions respectively (Stohr, 1974, p.11). In a similar framework, Pred talks about the role of large job-providing organisations in the growth and development of systems of cities:

> The large job-providing organizations that dominate advanced economies contribute to the growth and development of systems of cities in those economies in at least three inter-related significant ways (1) the generation of local and non-local multiplier effects; (2) the diffusion of growth-inducing or employment-creating innovations; and (3) the accumulation of operational decisions that affect the survival and scale of particular organizational units in particular cities. All three of these phenomena share the feature of contributing to the growth and development of systems of cities by ultimately influencing at what locations new jobs are created (Pred, 1974, p.9).

Underlying the notion of functional regions is the assumption that the region functions as a unit in order to maintain the existing intense inter- and intra-sub-system interaction that is essential for meeting a 'need' or 'needs'. Indeed the spatial redistricting of congressional districts after every census, in the United States, is an attempt at creating units that are 'ideal' for political processes and for sustaining the existing processes of the American democracy (See Morrill, 1973).

The fifth meaning of function is that associated with a gathering for a ceremony. For example, the religious functions in the locust cult region of China were gatherings where sacrifices were made to the locust gods. These religious functions have given rise to a spatial distribution of locust temples (Hsu, 1969, pp.731-52).

Functional Substitutability. If functional substitutability entails the substitution of a culturally alien object, the system, as we noted earlier, may not withstand the systemic ripples. For example, the rapid introduction of foreign technology into technologically less advanced societies creates new needs and new ecological balances (de Souza and Porter, 1974). For example, the introduction of snowmobiles and firearms to the Eskimos has resulted in a tipping of the delicate ecological balance between the Eskimo and the Arctic wildlife. Surely many would argue that the present international and internal upheavals in Iran may, in part, be due to the accelerated rate at which the Shah was Westernising the society. For similar reasons, David Davidson questions whether African countries can survive (Davidson, 1974). Geographic studies based on this premise of culture-specific functional substitutability can use the functionalist framework as a conceptualisation vehicle (see Szymon Chodak, 1973).

Goals. Implicit in the meaning of goals is action directed toward the achieving of a specific result. In geography, we have studied these goals in a spatial framework. For example, the goal of multi-national corporations is profit. However, in pursuit of this goal spatial patterns are generated and complex international linkages established. Similarly, the distribution of retail activities and office establishments in a capitalist Western city reflect the aggregate spatial manifestation of the individual entrepreneur's profit maximisation goals. Where the entrepreneurs fail to meet these goals the particular cluster of activities, such as a CBD, decline.

Pattern Maintenance/Homeostasis/Equilibrium. These concepts are central to studies of landscape morphology and to studies of human systems. For example, Shin-Yi Hsu used an equilibrium model to explain man's adjustment to a disaster (locust) which threatens to disrupt the *status quo* in parts of China (Hsu, 1969, pp.751-2). In a different context, Beaujeu-Garnier's study of France's attempts at decentralisation of population and resources from Paris, is another example (Beaujeu-Garnier, 1974, pp.113-25). Based on a policy of *amenagement de territoire* (or town and country planning), France has attempted to reduce the political, economic and social dominance of the Paris area by creating new regional foci. This process should, hopefully, create a balanced pattern of growth; 'an equilibrium not based mainly upon economic profits, but upon a better life for more people' (Beaujeu-Garnier, 1974, p.124).

Adaptation. It is the mechanism by which the system maintains that constancy which is essential for pattern maintenance. From the days of the Deterministic/Possibilistic debate, the importance of an organism's adaptation to its environment has been central to geographic thought. At a theoretical level, Hudson's study of the spread of settlement is an example of an adaptation study (Hudson, 1969, pp.365-81). Underlying his concept of fundamental niche is the assumption of organism-environment harmony.

Integration. In all formulations of functionalism, integration is a central concept. It implies: (i) a *status quo* for a long period of time; (ii) a group with norms and goals; (iii) a space occupied by the group; (iv) inter-sub-system interaction; and (v) movement. Subsumed in these is Ullman's 'prerequisites' for spatial integration – transferability, intervening opportunity and complementarity. In geography, the concept of integration has been widely used in studies dealing with the spatio-temporal associations between races and between ethnic groups, regional development and flows, and in the generation of functional or nodal regions.

Our aim above was to show that geographic systems include attributes of cultural systems. In fact, cultural systems are subsets of geographic systems. Both are examples of 'living systems'.

Direct Use of Functionalism in Geography. At both the empirical and theoretical levels, most of the geographic work involving functionalism have been in political geography. Because of the spatial focus of both political geographers and political scientists in the field of international relations, these attempts have followed either the political functionalism or the closely allied approach of Deutschian transactionalism. The broad areas of communality are:

(1) A stages approach – from tribes to nation-states and, finally,

to supernational organisations – is used.

(2) The centrifugality versus centripetality of Hartshorne captures the essence of the bi-polarities of the political scientists: conflict and co-operation, discord and collaboration, war and peace, to give some examples. The emphasis in functionalism is, however, on the centripetal forces – co-operation, collaboration, peace, et

(3) The search for the underlying spatial unit that maximises integration (intra-state or inter-state).

(4) The emphasis is on integration.

(5) That 'peace and security are to be guaranteed by the efficient provision of essential services to fulfil commonly-felt needs' (Taylor and Groom, 1974, p.2).

(6) Integration is spatial for the geographer but international for the political scientist.

(7) The systems concept is common to both political scientists and geographers.

(8) Related concepts of pluralism, federalism and dualism are used by both.

Most of these themes, whether direct or indirect, are re-echoed in the three classic papers on political integration. In Hartshorne's functional approach (Hartshorne, 1950, pp.95-130), differentiation in a political unit, say a nation-state, is the result of the functional organisation of the unit. Such functions fall into two counterpoised forces: centripetal forces which encourage integration and centrifugal forces which foster disintegration. Similarly, Gottman's circulatory theory of national integration includes most of the above points. In Gottman's conceptualisation, circulation (augmented by national iconographies) enhances integration (Gottman, 1951-2, pp.153-73). In Jones's unified field theory, the processes of spatial integration, after the creation of the nation-state, are largely restatements of the views expressed by Hartshorne and Gottman.

A direct application of Parsons' four functional prerequisites – pattern maintenance and tension management, goal attainment, adaptation and integration – is found in Bergman (1975, ch.2). He shows how these four requisites constitute a framework for political geography. At a more spatially specific level, Western Europe has provided a framework for such functionalist studies. Geoffrey Parker's book on *The Logic of Unity* (1974) and Johnson's recent paper on 'National Sovereignty and National Power in European Institutions' (1977), are examples in point. Parker's statement for European Community could easily have been attributed to Mitrany:

> Unity aims not only at countering national weakness, but also at bringing more positive advantages. It is obvious that a large economic unit must be more powerful than its small constituents, but it is also true that its economic efficiency can be far greater than just the sum of the parts (Parker, 1974, p.5).

Efficiency of scale is indeed an important feature of functionalism in a spatial context. This efficiency implies the potential loss of political power by the subunits and the emergence of super-national structures. An example in point is the European Parliament (see Johnston, 1977).

In spite of the above efforts, the functionalist approach is not buttressed in political geography. Rather, the emphasis has been on classification, areal differentiation and ideology. The 1969 observation of Kasperson and Minghi is largely true today: 'description, classification, and areal comparisons of formal spatial evidences of political structure have formed the core of political geography' (Kasperson and Minghi, 1969, p.70). Although the studies on political processes and voting behaviour are significant, they are not sufficient to invalidate that conclusion. In fact, Kasperson and Minghi, in the same paper, suggested functionalism as a framework for political geography because it encourages the study of features of the geographic landscape as a 'political system incorporating a complex set of relationships of behavior, process and structure' (Kasperson and Minghi, 1969, p.70).

The above discussion has demonstrated that geographers are interested in many of the themes prevalent in functionalism, and many geographic studies on a real functional organisation, ecology and agricultural regions, for example, do fit aspects of the functionalist framework. However, only a few geographers, particularly political geographers, have attempted to apply it systematically. Does this relative absence of functionalist studies imply that only politico-spatial systems do exhibit needs, have goals, have interacting entities and requisites?

Bases for a Functionalist Geography

To explore these bases, certain propositions are proposed and explored using a functionalist approach.

1. Geographic Systems can be Studied in a Systems

Framework. If we substitute 'geographical' for 'sociological' below, Parsons' views could be equally applicable to geography: 'It has seemed to me that many problems which have often been conceived to be primarily "sociological" could only be satisfactorily approached by attempting to place their social system aspects in the setting of the more comprehensive general action system' (Parsons, 1977, p.151).

In geography, systems analysis has increasingly become a vehicle of research ever since Ackerman espoused it in 1963 (Ackerman, 1963), in two ways: the analysis of physical and human systems and the use of the systems framework for analysis (Wilbanks and Symanski, 1968, pp.81-91). As Wilbanks and Symanski rightly noted, most geographic studies have tended to blend both viewpoints in the framework of either closed- or open-systems. In geomorphology, Chorley has specifically advocated the use of the open-system approach in contrast to the closed-system approach: 'open-system thinking . . . directs attention to the heterogeneity of spatial organization, to the creation of segregation, and to the increasingly hierarchical differentiation which often takes place with time. These latter features are, after all, hallmarks of social, as well as biological, evolution' (Chorley, 1962, p.301). As Eisenstadt and Curelaru noted, open-system models are also characteristic of functionalism (See Eisenstadt and Curelaru, 1976, pp.178-93).

In cultural geography, Foote and Greer-Wootten demonstrated, in their study of the Eskimo-environment system, that a systems approach could be used to understand culture-environment interactions. They noted also that 'each system can be located spatially and temporally by combining knowledge of environmental constraints and efficiences' (Foote and Greer-Wootten, 1968, p.89). For human geography, Langton equally espoused the virtues of systems theory. He asserts that system models of processes provide:

> the most powerful means of pursuing the causal analysis of change which have yet been devised . . . If the adoption of a systems approach had little effect beyond that of forcing us to consider seriously the problem of geographical entitation, then it would provide no mean service (Langton, 1974, p.171).

In a very elucidating book, Laszlo (1972) demonstrated how sociology has adopted a systems philosophy in the form of functionalism. Central to Laszlo's criteria for a systems philosophy are the existence of theory, the existence of systemic state properties of wholeness and order, the tendency toward self-stabilisation (equilibrium, pattern-maintaining, socio-cybernetic or spatio-cybernetic, etc.), adaptive self-organisation, and the existence of inter- and intra-group hierarchies. These attributes have been articulated in various forms by both cultural and human geographers (see, for example, Langton, 1974, Wilbanks and Symanski, 1968, Foote and Greer-Wootten, 1968, and Harvey, 1969). Geography, like sociology, is concerned with theoretical systems which are isomorphisms of the concrete systems; its focus is on order and pattern. Geographic systems exhibit tendencies toward equilibrium, pattern-maintenance, adaptive self-organisation, and, above all, have hierarchical structures.

The use of a systems framework for functionalism, does not imply that functionalism is just a sub-set of systems philosophy. General systems theory provides a methodological framework for certain functionalist studies, but functionalism itself is more associated with what to investigate, in what order, and why. Functionalism provides a framework for 'purposive' systems studies that revolve around the basic functionalist concepts discussed earlier in this chapter. A complex 'flow chart' or a system of mathematical equations that capture the 'systemness' of a particular geographic topic do not, by themselves, make that study a functionalist study. Functionalist philosophy emphasises the needs, goals, roles and the linkages in the systems.

2. Spatial Behaviour is Goal-Oriented. At the disaggregate level, human behaviour is not random but purposive (Tolman, 1932); this is especially true of spatial behaviours such as shopping, work, leisure and vacation. Implicitly, therefore, spatial behaviour is goal-oriented. At the aggregate level, broad goal-oriented behaviour is generally assumed in systems studies. This is true of the Eskimo fishermen (Foote and Greer-Wootten, 1968), the actors in rural-urban migration (Mabogunje, 1970), and the buyers in a housing market. These aggregate goal-oriented behaviours give rise to spatial orders which are manifested in synchronic and diachronic patterns. The synchronic, static, morphological 'snapshots' of Sauer, whether they are related to house types, agricultural patterns, the distribution of towns of varying sizes, or to coal mines in Appalachia, reflect aggregate purposive goal-oriented behaviour. Similarly, diachronic patterns which give rise to dynamic order or spatial processes, also reflect goal-oriented behaviour. This assumption of order, either social or spatial, is central to both sociology and geography. To quote Laszlo:

> If social systems were unordered, in time and space . . . social [and
> spatial] theorists would be creating imaginary orders when discussing
> morphological and dynamic properties and would have to conclude
> that such schemes are inapplicable to actual social [and spatial]
> systems . . . There can be no science of a phenomenon in a constant
> state of flux (Laszlo, 1972, p.101. Insert in parenthesis is author's).

3. Feedback Loops are Important in Spatial Systems. Goal-seeking
behaviour and the resultant pattern are also related to the occurrence
of feedback loops. In functionalism, feedback is indeed an important
behaviour-regulating process. For example, in his discussion of
leadership as social exchange, Merton underscores the importance of
negative and positive feedback cues to the regulation of the individual
leader's action (Merton, 1976, p.79). This relationship between
goal-seeking and feedback in human geography has been noted by
Langton:

> Population growth stimulates industry which stimulates further
> population growth; under increasing returns to scale, manufacturing
> stimulates mining which stimulates further manufacturing; trade
> stimulates further manufacturing; trade stimulates production which
> stimulates further trade, and so on. Positive feedback is the
> mechanism responsible for the growth, increasing specialisation and
> differentiation which are characteristic of all living systems
> (Langton, 1974, p.145).

4. The Components of a Spatial System Fulfil Functional Requisites.
The assumption of goals, adaptation and integration implies that spatial
and social systems have requisites that are designed for specific goals,
and each sub-system has a specific function. At the risk of illegitimate
teleology, we could generalise that central places exist to provide
central functions, industrial systems to generate profit, and
transportation systems to move people and goods. In the last
sentence, 'exist' implies only necessary conditions. However, since
central places, industrial systems and transportation systems may be
regarded as 'self-persisting systems', they may be amenable to
functional explanation (See Harvey, 1969, p.436).

Related to the question of requisites is that of 'needs'. Spatially, it
could be argued that the satisfaction of social, economic and political

needs generate spatial needs that structure the nature of the resultant space. Let me elaborate. In Western Europe, after the demise of the Empires and after the industrial devastation of the Second World War, the need to counter national weakness and the need to create a large, viable economic system, *inter alia*, implied the need for enhanced mobility of labour, capital and technology. These needs were directly related to the need for spatial integration, but above all, they resulted in the spatial reorganisation of power, industries and transportation, not for the benefit of a single nation, but for the common goal of the European Community.

5. Spatial Unit is One of Functional Unity. 'If geography is concerned with space, then it is also concerned with region. We cannot survive as a contemporary discipline if we are intellectually unable to delimit the systems we study' (Mayfield and English, 1972, p.425). The regions we derive and study exhibit unity on some set of subjective criteria. Indeed Vidal de la Blache was right when he suggested that 'the phenomena of human geography are related to terrestrial unity by means of which alone they can be explained' (quoted from Harvey, 1969, p.441). Terrestrial unity is nothing more than functional unity between complementary physical and human sub-systems within a given space or region. This unity underlies Hartshorne's definition of a region as:

> at least a loose unity, in the sense of a complex of related elements forming a relatively closed system . . . Only if this be true can we properly say that the region – as distinct from things within it – 'has structure, form and function and hence position in a system' (Hartshorne, 1961, p.276).

The regions we study vary in scale, but they all reflect some functional unity; a unity which Penck (see Hartshorne, 1961, p.265) has characterised as *Gestalt* because the region is indeed a 'dynamic structure in which the parts reach into each other functionally and can be understood only in view of the whole' (Hartshorne, 1961, p.266). If a region is a unity based on *Gestalt*, it implies a wholeness based on the non-summativity of its components. Attributing systemic properties of *wholeness* and *order*, implies a functional unity analogous to those of cultural units; it also requires general systems theory which, in the words of Bertalanffy, 'is the scientific exploration of "wholes" and "wholeness" . . . Hierarchic structure, stability, teleology, differentiation,

approach to and maintenance of steady states, goal-directedness –
these are a few of such general system properties' (Laszlo, 1972,
p.xviii).

Summary and Conclusions

This review of functionalism was largely designed to expose the student
to a viewpoint that has gained some degree of acceptance in many social
science disciplines. It has certain attributes that have made it appealing.
Although not in wide use in geography, we believe, and we have
demonstrated, that many studies in geography are indeed examples of
functional analysis. Presently our discipline is experiencing an explosion
of systems studies that do, in many ways, have traces of functionalism,
but are not clearly distinguishable by any philosophical or
methodological distinctiveness. Geographic functionalism may create
such a canopy. Even if it does not provide us with a concise coherent
philosophy of the world, it does provide a refreshingly social science-
like viewpoint for explaining the complex relationships between man
and his environment in a functional systemic context. To quote David
Harvey, 'it provides us, rather, with a convenient umbrella term with
which to characterise rather varied viewpoints that have something in
common' (Harvey, 1969, p.439).

After reading other chapters in this book, the student may wonder
whether functionalism is indeed a philosophy. From one view point, it
is a form of teleological philosophy that explains situations from a
cataloguing of the roles. From another viewpoint, it is a type of
systems philosophy (Laszlo, 1972) with a strong organismic emphasis
which stresses the interrelationships between sub-systems; an approach
and a viewpoint that can be geographic. Much work is needed and this
paper was designed to initiate such a move rather than to articulate
it fully.

5   PHENOMENOLOGY

Edward C. Relph

Introduction

'Positivism . . . decapitates philosophy' declared Edmund Husserl (1970, p.9), the originator of phenomenology, and he meant that in its single-minded addiction to facts positivism omits all the higher and broader questions of life. In the first years of this century Husserl reacted against the then fashionable attempt to reduce philosophy to a set of psychological generalisations. He argued that philosophy was not a factual science, was not grounded in factual science and could not use the methods of factual science, and he began to formulate a way of thinking about philosophical questions that did not make assumptions about what could be studied and how, but which was responsive to all phenomena of human experience. The first accounts of this method of phenomenology were directed to issues of academic philosophy, but as his thinking developed Husserl came to realise, as perhaps Blake, Ruskin, Kierkegaard and others had before him, that the reductive and decapitating methods of positivistic thinking, which accept as true only that which can be objectively established, had pervaded many aspects of human existence for which they were inappropriate. Belief in art, religion and existence was being actively undermined. 'Merely fact minded sciences made merely fact minded people' Husserl exclaimed in his last work, *The Crisis of the European Sciences* (1970, p.6), for when the methods of scepticism and detachment that have proven so successful in natural sciences are applied to human concerns they reduce our lives to mere facts or accidents of nature. In contrast, phenomenology revives the issues of ethics, knowledge and being that have been the traditional concerns of philosophers, and it provides a means of exploring all facets of human experience without rendering them devoid of meaning.

By the time of Husserl's death in 1938, phenomenology had achieved acceptance not only in philosophy, but also in anthropology, sociology and psychology, and it had influenced the work of poets, artists and novelists. This acceptance has since grown to the point where phenomenology is a widely adopted approach in most of the social sciences. In geography, however, the impact of this method has been limited to a handful of methodological statements and substantive investigations, and it is clear that among geographers there is little familiarity with phenomenology. Indeed, because of the difficult and unusual language used in some of the methodological papers, there is probably much confusion. I do not wish to exacerbate this confusion by writing yet another account of the technicalities of phenomenology, so my aim here is to provide a clear and relatively non-technical circumscription of this approach, as well as to survey phenomenological writings by geographers. In doing this I am sure I will fail to consider concepts and authors others think to be essential, and I can only plead that I am concerned less with precise assessment than with giving a sense of the insights phenomenology has to offer as a way of thinking geography.

Initial Circumscriptions

For those predisposed to dismiss phenomenology as a non-rigorous and overly introspective method I must affirm at the outset that it is a coherent and complete philosophical position from which to address the world. It acknowledges its antecedents and relations to other philosophical traditions, offers critiques of their arguments, and provides its own alternative arguments. Therefore, criticisms or discussions which treat phenomenology as though it is a technique somehow comparable with mental mapping or multivariate analysis are not justified. The breadth and depth of phenomenology are substantial and radical; they can be well illustrated by drawing a contrast with the philosophy of science.

The philosophy of science, as apparent in scientific method, is a way of thinking that constitutes itself as *a way of doing*. I mean by this that it tells us how to organise concepts and facts by means of experiments and tests that will provide explicit results. In verifying hypotheses it is necessary for us to separate ourselves from the structures and processes of matter or of society and to treat them as external to ourselves, thus demonstrating control over the subject matter whether atoms, rats in mazes, or people spatially distributed. Indeed if such control does not exist the experiments cannot be replicated and hence must be scientifically invalid. Furthermore, because of its detachment this way of thinking conceives itself as non-social and non-political, that is, as value-free. Probably everyone knows that this is a pretense, that values and ideologies cannot be held in abeyance; but the game of detachment in research is played even as science works through missiles, television sets, automobiles and city planning to change our lives in countless ways.

If scientific method is a way of thinking that realises itself as a way of doing, phenomenology is a way of thinking that reveals itself in *a way of being*. Rather than treating the world as somehow independent of us, it requires that we reflect on our own consciousness of things and explore the various manifestations of them in our experiences and in so doing come to a deeper understanding of ourselves. Subjective encounter, the very attitude that is disposed of in scientific method, is taken to be fundamental in phenomenology, for it is only through my consciousness or experience of things that I can possibly know them. In other words phenomenology insists that I must think about how I experience the world; my dispassionate assessments, frustrations, imaginings and emotional outpourings are all considered to be facts of experience, and one fragment of my experience is not arrogated into some special status. In fact, quite the opposite is required; I have to learn not to see objective measurable facts as somehow more important than my own experiences.

To clarify what this means consider the phenomenon of time. From a scientific perspective time is unidirectional, a constant frame of reference; our lives happen in the context of, yet apart from, time. Phenomenologically time is part of our living; it is variable and multidirectional for we have memories and moments of prescience; there are days which hang heavily and weeks which fly past. From this point of view it makes sense to say of somewhere nerve-racking and hideous, like London's Heathrow Airport, that 'I spent several days there one afternoon'. A phenomenological study begins with such experiences and attempts to sort them out by identifying varieties of time experience and the consistencies within them.

The consequences of the exploration of our own consciousness are profound. Husserl (1970, p.137) wrote '. . . the total phenomenological attitude is destined to effect a complete personal transformation comparable in the beginning to a religious conversion . . .' For geographers imbued with impersonal methods based on statistical surveys this is no doubt mystical and heady stuff. But for anyone with a serious commitment to their work it should not be difficult to grasp. It means that while phenomenology is a formal method of thinking and seeing, this method leads to insights that once attained cannot be dismissed at will and which will change our attitude to existence.

If phenomenology effects such profound changes in the outlook of individuals it follows that it must have tremendous social and political consequences. It is, however, not clear what form these might take for they have never been realised. Phenomenology remains a peripheral way of thinking in a society governed by science and technology.

Varieties of Phenomenology

Since phenomenology was initially an approach to philosophical problems it is to be expected that much phenomenological writing has been about issues in academic philosophy. The importance of this work can scarcely be overestimated since it provides a firm foundation for other studies, but it is the case that much philosophical phenomenology is expressed in a technical and obscure language. From a strictly philosophical perspective the development of such a language is essential in order to differentiate meanings and concepts, but to the outsider it is exceedingly difficult to penetrate; the language with its innuendoes and subtleties has to be learnt just as one has to learn the rules of calculus. So rather than plunging directly into the works of Husserl or Heidegger or Merleau-Ponty, the phenomenological neophyte is better advised to start with the writing of some of their interpreters such as John Wild (1963), Don Ihde (1977), Maurice Natanson (1963) or Stewart and Mickunas (1974).

As a philosophical approach phenomenology is a radical way of examining the phenomena of our consciousness or experience (the two words are virtually interchangeable in phenomenology). The stock in trade of philosophers is thought and discourse – how to think logically, dialectically or whatever – and while for most geographers method involves actions such as analysing data, making maps or conducting experiments, for philosophers method has to do with ways of thinking about problems. So when it is said that phenomenology is 'a presuppositionless return to the things themselves', this refers to how we are to think about things. In particular it proposes that rather than analysing or developing the opinions of philosophers or other geographers, we should be concerned with recapturing our sense of wonder about the world we live in and with trying to forget the concepts and abstractions we have acquired in our formal education and to reflect on what we encounter directly. Since our experiences are as diverse as life itself, the potential subject matter of phenomenology is the entire realm of human experience. In philosophy, however, phenomenology has been used to examine phenomena of philosophical interest such as existence, time, religious encounters, modes of thinking and being.

A second type of phenomenology is that used in academic disciplines other than philosophy. In these, phenomenology has usually been employed in a less radical manner, and rather than being taken as a complete reorientation it is accepted as one perspective among several alternatives, such as positivism, structuralism and Marxism, that can provide a basis for studying social issues. There are, of course, many topics that cannot or need not be examined phenomenologically, like foreign aid transfers, commuting patterns or the history of cities. But because it accepts the complexities and meanings of human experience, phenomenology is especially appropriate for studying how individuals relate to each other or to their environments, for considering the needs and satisfactions of people or for investigating any topic that touches on the issues of existence.

Possibly the strongest statement for phenomenology as an academic method is that formulated by Alfred Schutz (1962), a German sociologist who wrote during the 1940s and 1950s. He argued, and indeed demonstrated in his own writings, that phenomenology is empirical because it is based on observation, it is systematic because it is concerned with the organisation of the phenomena of experience, and it is rigorous because it is reflexive, that is, it subjects its own procedures to critical appraisal. In fact, Schutz suggested that by using phenomenological methods it is possible to formulate propositions that can be tested and verified. This is probably leaning too heavily on the language of science for many phenomenologists, and regardless of whether phenomenological observations are verifiable the difference from scientific method always remains – phenomenology is directed to the acceptance and full appreciation of the facts of experience, not their reduction into elegantly simple explanations.

A third type of phenomenology is suggested in Husserl's statement that the total phenomenological attitude is tantamount to religious conversion. Much of the writing of phenomenologists is not just scholarly and academic reflections about existence, but is a call to a way of living. Phenomenology recognises that the 'most real world' we live in is that of fellow human beings and commonplace objects; without them life would be lonely indeed and scarcely human at all (Schumacher, 1977, p.24). It leads us back to the ways in which we relate to others and the meanings we struggle to see in our own existence.

Although difficult to grasp, Martin Heidegger's writing has been the most influential contribution to this 'lived' or existential phenomenology. His early work was a complex study of the phenomenology of being, but his later investigations became increasingly poetic and spiritual as he grappled with the implications of his radical insights into the nature of human existence (Heidegger, 1971). His work has been a source of inspiration for theologians who have attempted to show that by adopting phenomenological methods of thinking and seeing we can demystify religion and make it a part of immediate human experience once more.

This effort recalls John Ruskin's comment (1905) that 'To see clearly is poetry, prophecy and religion – all in one.' Phenomenology is an attempt to see clearly. If we can achieve such vision it is bound to change our attitudes to our own existence, to other people and to the world around us. It will make us more aware of the meanings of our lives, more critical of theories and ideologies which restrict the quality and variety of existence, and, to use a term of Heidegger's more inclined to 'spare', that is, to avoid imposing our wills on things, environments or other people. This attitude of 'touching the least' requires that we develop a sense of caring and responsibility for things, places and people that allows them to be themselves with the minimum interference from us. This is, I think, the essence of Heidegger's existential phenomenology, and if we can accept it and put it into practice rather than merely talking about it and hypocritically acting otherwise, it must have profound implications for the manner of our living.

Phenomenological Method – The Things Themselves

Phenomenology discloses and elucidates what we experience and how we experience it. This involves probing behind what we take for granted in everyday living and thinking. There are many styles of doing this and it is misleading to speak of phenomenological method as a rigid set of procedures, though some authors have identified key phases in this method and one, Don Ihde (1977), has gone so far as to suggest a sequence of rules which can be followed in order to break down habits of thinking and to develop phenomenological insights. Such rules are, however, never more than rough guidelines for beginning to think phenomenologically. I believe it will be less misleading if I discuss phenomenological method without reference to rules and stages, but instead in terms of the crucial notion of 'the things themselves' to which phenomenology is meant to return.

Our experiences of things, whether illusory or substantial, transitory or enduring, are taken as the given facts which phenomenology explores. This statement needs elucidation, so let me offer an illustration. A few years ago William Kennedy was interviewing the novelist Gabriel Garcia Marques in Barcelona (Kennedy, 1973, pp.57-8). They were walking in the city centre when Kennedy saw a trolley car cross the street they were on about three blocks away. It crossed from right to left, was visible for only a second or two, but he could see that it was yellow and old fashioned. No one else noticed it. Now what was interesting was that there had been no yellow trolleys in Barcelona for several years; the last one had been buried in a formal ceremony two years earlier. Kennedy writes: 'What had I seen? I have no idea. "To me", Garcia said, "this is completely natural." ' A scientific explanation of this experience would begin sceptically and look for a yellow truck that had crossed the street, or perhaps explain it away as some sort of hallucination and so deny that the experience was real. On the other hand a phenomenological account would accept this experience just as it had happened, as Garcia did, and consider its various implications and subtleties, its context and significance.

It is the investigation and description of the world as we experience it directly and immediately that is attempted in phenomenology. Such exploration requires that we break down ingrained habits of seeing and thinking and attempt to see things as though for the first time, with as little prejudice and as few assumptions about them as possible. Numerous artists have attempted something similar: Goethe (1970) tried to see 'with clear fresh eyes' and to abandon his old habits of thinking when he undertook his pilgrimage to Italy; Alberto Giacommetti (1964) described his way of experiencing the world as one of constant wonder and surprise. Such a suspension of judgement may sound naively simple but it is likely to involve considerable tension and the overthrow of cherished beliefs. E. F. Schumacher (1977, p.44) writes of the problems of developing self-awareness in *A Guide For the Perplexed:*

There is nothing more difficult than to become critically aware of the presuppositions of one's thought. Everything can be seen directly except the eye through which we see. Every thought can be scrutinized directly except the thought by which we scrutinize. A special effort, an effort of self-awareness is needed: that almost impossible feat of thought recoiling upon itself – almost impossible but not quite. In fact this is the power that makes man human . . .

Such concerted efforts for self-awareness demonstrate that
phenomenology is not merely introspective, for what is involved is
is not a casual consideration of how I feel, but a systematic reflection
on the character of my own experiences. Furthermore these are not
locked up within each of us, but are intersubjective because, as Doris
Lessing (1973, pp.13-14) has argued:

> . . . nothing is personal in the sense that it is uniquely one's own.
> Writing about oneself, one is writing about others since your
> problems, pains, pleasures, emotions – and your extraordinary
> and remarkable ideas – can't be yours alone . . . growing up is after
> all only the understanding that one's unique and incredible
> experience is what everyone shares.

There has been much discussion of the further stages of
phenomenological method: how to maintain awareness, how to
identify order and structure and meaning in the chiaroscuro of
experiences. Such discussion cannot mean a great deal until this
first effort at self-awareness has been made, and Ihde (1977, p.14) has
observed that without doing phenomenology it may be impossible to
understand it. If this seems too subjective consider that much the same
can be said of mathematics, of speaking a second language, of reading.
The challenges and insights that phenomenology offers come from
using this way of thinking and no amount of writing by me can convey
them adequately. Suffice it to say that in practice a phenomenological
investigation responds to the phenomenon and that the end result
is a description or interpretation that should increase our appreciation
and understanding of that phenomenon of experience. Like any
interpretation this can be revised and improved on my making more
sophisticated observations, adducing more evidence or having deeper
insights.

Geography and Phenomenology

In the numerous geographical periodicals and books published since
1970 there are, to my knowledge, just nine papers which deal explicitly
with the relations between geography and phenomenology, and six
books which use phenomenological approaches to examine
geographical phenomena. Most of the methodological papers present
similar arguments to those made here so there is little to be gained by

reviewing them in detail. However, it will be helpful if I outline their genealogy before assessing the substantive phenomenological contributions, for it seems that the former are concerned chiefly with academic phenomenology and the latter are in the rather different tradition of existential phenomenology.

The genealogy of the methodological papers is, I think, as follows. Husserl formulated a philosophical approach that countered scientism in philosophy and this was adapted for the study of social issues by Alfred Schutz (1962). Mercer and Powell (1972), in their examination of alternative approaches to positivism that might be relevant for geography, took the phenomenological writings of Schutz and stressed the possible basis for a human geography more open to subjectivity and meaning. Their proposal seems to have influenced Walmsley's (1974) rather more specific suggestion that phenomenology might prove to be especially useful in historical and behavioural research in geography. Entrikin (1977) reviewed the writing by geographers that he considered to be humanist – most of which had a phenomenological inspiration – but concluded in opposition to Mercer and Powell that phenomenology does not constitute a significant alternative to scientific approaches in geography. He argued instead that it has a role as a way of criticising those approaches.

In his paper on negativism and historical geography, Billinge (1977) returned to the writings of Husserl and to the proposal of Mercer and Powell, assessed phenomenological statements by geogiaphers and concluded that there is little in them that is 'phenomenological at its hardest core'. He also implied that even soft-core phenomenology has little to offer for 'historico-geographical' work. The philosophy of Husserl, and also the sociology of Schutz, influenced Buttimer (1976) and Ley (1977) in their consideration of the relevance of phenomenological approaches for social geography. They both consider that phenomenology would lead to broader and more sensitive social geographies in which elements of scientific enquiry could be combined with a humanist and phenomenological orientation.

Developing his ideas from the papers by Entrikin, Billinge and Buttimer, and again drawing directly on Husserl and Schutz, Derek Gregory in *Ideology, Science and Human Geography* (1978) makes a complex case for phenomenology within a critical theoretical geography. This is a concept of geography as a scientific discipline, yet one which recognises the ideological bases of knowledge and therefore requires its practitioners to be self-critical and aware of the grounds for the explanations they make. The role of phenomenology in this comes from its reflexivity, or self-awareness in making interpretations. Gregory stresses reflexivity within the context of committed explanations of geographical processes and structures, that is, explanations with a clear social and political commitment. Consequently his account of phenomenology does not dwell on the meaning of individual experiences, nor on the phenomena of experience, but emphasises method.

This summary discloses, even to a partisan commentator like myself, a certain confusion and an inclination to accommodate phenomenology to the prevailing ethos of science. The sequence of arguments seems to have gone like this: Phenomenology is a good alternative to scientific methods; at least, it looks useful for some types of geography; well, it could be a valuable form of criticism; perhaps it can be used to complement the work of scientific geographers from a humanist perspective; ah yes! phenomenology provides some significant qualifications within a committed scientific approach to geography. These authors appear to be seeking a more sensitive and humanistic approach for human geography, but they also want to apply it to conventional problems and without questioning the apparent gains that have resulted from the recent shift to a more rigorously scientific geography. Their interpretations are insufficiently radical; phenomenology is indeed a form of criticism, but of the very foundations of scientistic thinking rather than as a mere worrying out of the details of scientific procedure in the hope that this will somehow make science more humanistic. And it redefines subject areas and identifies new issues for study instead of being just a different technique for investigating traditional topics. Furthermore there have been no substantive studies in geography using this type of academic phenomenological method, and one gets the impression that, in all the writing about how to do phenomenology but never actually applying it, the arguments have quietly been absorbed by the conventional wisdoms of scientific geography.

A more radical type of phenomenological geography, and one which has led to substantive investigations of geographical phenomena, corresponds with lived or existential phenomenology in which methodological matters are implicit rather than explicit. In his paper on 'Geography, Phenomenology and the Study of Human Nature', Yi-Fu Tuan (1971) explores some of the relations between Heidegger's phenomenology and geography, especially the notion of man-in-the-world, that is, the individual as implicated in and responding to places and landscapes he encounters in his everyday living. Tuan argues that a phenomenological perspective can make us more aware of our spatial and geographical experiences, such as those of 'home' and 'journey', and more sensitive to the meanings of our relationships with environments. From this we begin to understand that phenomenology is not a method that can be applied simply to existing geographical topics, but that it restructures subject matter and leads us into discussions of such things as the relations between body asymmetry and spatial asymmetry, or egocentric and ethnocentric space – topics which are perhaps new and strange for geographers as academics, but which should be familiar to them through their encounters with the geographical life-world.

The Geographical Life-World and Phenomenological Geographies

Phenomenology has to do with beginnings, with phenomena that are first lived and experienced and are only subsequently formulated as concepts. Geography as a formal body of knowledge presupposes our geographical experiences of the world. In other words, geography has an experiential or phenomenological foundation (Relph, 1976a). Concepts of space, landscape, city, region, have meaning for us because we can refer them to our direct experiences of these phenomena. We live in a world of buildings, streets, sunshine and rainfall and other people with all their sufferings and joys, and we know intersubjectively the meanings of these things and events. This pre-intellectual world, or life-world, we experience not as a set of objects somehow apart from us and fixed in time and space, but as a set of meaningful and dynamic relations. That is to say, other people, objects, types of scenery, architecture and places matter to us to greater and lesser extents; we are concerned about them and we care for them.

The geographical life-world is a part of the total life-world, not clearly separable of course, but identifiable through a set of specific interests that have to do with places and environments rather than other individuals or plants or whatever. It is both social and natural, consisting of the pedestrians we notice when driving down a street, the scenery we see when hiking in the country, the clouds, the litter in the gutter, trees in blossom and indeed everything we see and sense out of doors. It is the world experienced as scenery, as an everpresent backdrop to our lives; but it is also the constant and unavoidable context of our lives, affecting our activities and intruding into our thoughts in countless ways.

Our experiences of the geographical life-world have been investigated by Eric Dardel (1952). He uses the term 'geographicité' to describe our pre-reflective or pre-conceptual experiences of space, place and landscape, a word which can be translated as 'geographicality'. Geographicality binds people to their surroundings, manifesting itself in a sense of place and in sensitivity to landscape. Everyone of us has to participate in the creation, maintenance and destruction of landscapes in order to stay alive; in such activities and in admiring sunsets or feeling overwhelmed by the pace of a city's downtown, geographicality is central. Dardel (1952, p.47) write: 'Geographical reality demands an involvement of the individual in his emotions, his body, his habits, that is so total that he comes to forget it just as he forgets his own physiological life.' It is accepted, taken for granted, and it is always specific. Our geographical experiences are of *this* street, *this* valley, *this* place.

All the substantive phenomenological investigations by geographers have been either of phenomena of the geographical life-world or of particular aspects of the experiences of geographicality. In *Topophilia*, Yi-Fu Tuan (1974) examined the variety of pleasant experiences of landscapes and places. Topophilia is a home-directed sentiment, one that is comforting and relaxing, but it can also be an ecstatic and uplifting experience of either a natural or a man-made setting. Tuan's study of this experience is implicitly phenomenological, that is to say his attention to the variety and meanings of topophilia experiences are clearly phenomenologically inspired, but he nowhere discusses his approach. I think this is so for two reasons. One is that the subject matter of environmental experience and the phenomenological method by which it is being investigated are not different but merge one into the other. The other may be that Tuan did not wish to be tied by methodological restrictions, for he considered an accuracy of principle and style to be more important than a rigid following of methodological rules. Tuan's method is one which draws carefully on a wide variety of accounts of environmental experience and offers a coherent interpretation of them. So rather than examining his own environmental experiences directly he approaches them indirectly through the sensitivities of others.

A quite different approach to experiences of the geographical life-world is that used by David Seamon (1979) in his investigation of movement, rest and encounter. His interpretations are based on reports of daily environmental experiences he solicited from a number of individuals. The reports were generated through group discussions of topics such as the focal places in the participants' lives, what they noticed in particular environments, how they decided on destinations for trips, whether they always followed the same route to work, what parts of environments they ignored, what manifestations of spring they noticed and how they reacted to them. It is scarcely possible to do justice to this approach here, but it should be noted that the participants were given very little direction and were actively involved in a process of increasing their self-awareness of environmental experiences; they were not treated as convenient sources of information. The reports provided a wealth of specific and individual reflections about environmental encounters which Seamon structured in terms of three themes: movement has to do with the habitual nature of much of our movement, whether crossing a room or crossing a city; rest refers to the significance of belonging to somewhere, having a place where we can regenerate our energies; and encounters are those situations of heightened consciousness when we are especially aware of environments.

Rather than stressing experiences, phenomenological studies can emphasise the phenomena of the geographical life-world, and there have been two such studies, both having to do with space, place and sense of place. In *Place and Placelessness* (Relph, 1976b) I employed phenomenological procedures and concepts to examine the spatial and experiential contexts of places. Instead of examining my own experiences of places directly I tried to argue that cultural meanings and attitudes are made manifest in landscapes and that a phenomenological investigation of landscapes can reveal the character of the underlying geographicality. Modern landscapes appear to show that scientism and technique have reduced our abilities to experience or to create significant places.

In *Space and Place* Yi-Fu Tuan (1977) again used phenomenology implicitly. In this book, as in *Topophilia*, he proceeds from physiological experience, through cultural aspects of space, to the increasingly complex and subtle forms of sentiment and feeling about places. Again he draws his material from the diverse observations of others, but one supposes that he is using these to grasp the character of his own experiences of space and place.

Doing Phenomenological Geography

A recurrent theme in the phenomenological writing of geographers is that phenomenology leads to self-awareness and a heightened sense of responsibility for the environments we live in and the ideas we express about them. Phenomenology helps people to appreciate their worlds and lives more fully and also to begin to see what modern science, technology and consumerism have wrought in the name of efficiency and satisfaction. So it might be suggested that the major role for phenomenology lies not in research but in teaching, and that once the insight of grasping the world from the perspective of one's own experience has been achieved then necessarily one must do geography phenomenologically. Of course it takes time and effort to translate this insight into consistent practice, but in the meantime motivations and interpretations will fall towards the human sentiments of phenomenology and away from the abstract strictures of scientism.

At least, I would like to think that phenomenology can be communicated thus directly. In fact it may be that the conceptually preordered ways of scientistic thinking have become so deeply entrenched that even for someone quite disenchanted with them any work which does not follow accepted patterns of data collection, analysis, model building and so on is likely to seem second rate. Moreover, phenomenological studies will be criticised by the scientifically minded as trivial, lacking in objectivity and impractical because they have no direct policy implications. In order to do sound phenomenological investigations in geography it is necessary to transcend these self doubts and criticisms. This is difficult because there is no external framework on which you can depend and which can take the responsibility out of thinking. In a phenomenological study if the end result is poor it is because your thinking is weak and you cannot shift blame to the lack of data or inadequate theory.

These are strong words and the fact is that the few phenomenological studies in geography have tried to evade the full measure of this responsibility by leaning on the accounts of authorities or other people's experiences, while several of the methodological accounts have hedged about the radical differences between scientific and phenomenological methods. In short, phenomenological writing by geographers has been cautious and lacking in self-confidence. I do not intend a counsel of perfection by this; the measure of any work is whether it adds to our knowledge of the world or our understanding of ourselves or enhances life, not whether it follows methodological rules precisely. But I do believe that if phenomenological geographers do not constantly keep in mind the philosophical origins of phenomenology as a radical critique of positivism and scientism, then we are likely to find ourselves participating in some mongrel methodology that gives us numerical models of topophilia or experiential expositions of urban systems. Or, more probably, the profound methods and insights of phenomenology will slowly but surely be subverted by the dominant scientific modes of thinking and doing.

If such subversion can be avoided the potential subject matter for phenomenological geographies is enormous, though different from most conventional topics. Research subjects are generally defined by tradition or by practical and policy concerns, and it is not possible to study urban systems, filters in housing markets or decision making in resource management from a phenomenological perspective. Phenomenological methods and topics define each other so that subjects must be either phenomena of the geographical life-world, for instance, streets, particular places, suburban landscapes, expressways, mountains and plains, or they may concentrate on experiences such as obliviousness to some environments and engagement with others or the aesthetic and spiritual sentiments engendered by landscapes.

These types of topics may seem frivolous and impractical since they have no direct or obvious planning implications. Whatever the arguments for or against applied geography the case is that it is fashionable to use geographical knowledge to make a case for supposed improvements in social and environmental management. So one question that is bound to be put by someone is whether phenomenology can be used in planning. Such a question reveals the depths to which scientism has penetrated, for it implies that in our society the only sensible way of organising people or modifying environments is by the application of formal expertise and technique. It is as though people did not think carefully, organise communities or build towns before the advent of rational science.

There are, of course, many ways of thinking and doing: religious, vernacular, emotional, rational, premeditated, spontaneous, habitual and so on, all of which can be employed carelessly or carefully, for good results or poor results. Scientific and rationalistic ways have somehow been arrogated into a position where they dominate all the others. Phenomenology aims at reinstating the other ways of thinking and doing, and such a radical goal means that it has little of immediate practical value to offer society. Yet because it can change the outlook and being of individuals, the long-term significance of phenomenology for the way in which people create their own worlds could be considerable. Exactly what form these worlds would take I have no idea, but in moments of utopian speculation I wonder about the implications of Kropotkin's notion of *Mutual Aid* (1972), or perhaps a little more substantially about suggestions for increased local autonomy and the development of intermediate technologies. Schumacher (1977) has argued convincingly that local autonomy and ecological stability without a metaphysical or philosophical reconstruction to support them will be empty achievements. Phenomenology has a central role to play in a metaphysical reconstruction which rejects a dependence on economic efficiency and material satisfaction for a way of life that acknowledges and enhances all aspects of human existence.

6 AN EXISTENTIAL GEOGRAPHY

Marwyn S. Samuels

Introduction

There is a sentiment shared widely among contemporary social critics and scientists, as well as among various human geographers, that we are now in the midst of a burgeoning 'crisis of faith' in the hitherto sacrosanct domain of scientific method and philosphy (Ravetz, 1973; Wail, 1965; Zelinsky, 1975; Buttimer, 1974; Relph, 1970; Tuan, 1971; Mercer and Powell, 1972). Dissatisfaction with positivist epistemology and goals in the study of man has raised the spectre of various new frontiers in social thought and theory. Some of these new frontiers have emerged to challenge and even abandon the spirit of objective, quantitative and deterministic analysis. Their languages often speak with a humanistic syntax that urges a concern for human value, quality, subjectivity, and even spirituality. Some appear thoroughly idealistic and inclined toward the solipsistic.

Despite that appearance, however, not all such new frontiers are idealistic and subjectivist. Marxism (see Chapter 10) and existentialism, for example, are both poised to attack idealism in almost any form, and especially in the form of subjective idealism. Existentialism is, furthermore, most easily summarised as an endeavour, on the one hand, to restore the concrete, immediate experience of existence *in situ* to the realm of knowledge, and on the other hand, to bridge the logical gap that separates subjective and objective, idealism and materialism, and essence and existence. Such is the central message of Jean Paul Sartre's famous phrase 'existence comes before essence' (Sartre, 1946; quotation from a translation by Novack, 1966, p.74).

There are, no doubt, many different ways to interpret and analyse the intent of that formula. We could follow Sartre's own explanation to the effect that the phrase 'existence comes before essence' means that 'man first of all exists, encounters himself, surges up in the world – and defines himself afterwards'. This, as we would see, further means that to understand man we must first of all 'begin with the subjective', for man is 'a kind of project that possesses a subjective life', and 'man is nothing else but that which he makes of himself'. This, in turn, would lead us to the first principle of existentialism which is that 'once thrown into the world, [man] is responsible for everything he does'. Existence comes before essence, as it were, because man is free (Sartre, 1946; Novack, 1966).

The methodological insistence to 'begin with the subjective', and the principle of existential freedom, however, require elaborate analysis, for neither means what each appears to mean at first glance. To 'begin with the subjective', for example, is not an incitement to indulge the *ego*. On the contrary, it is to insist that the human subject is firmly grounded in irreducibly concrete historical and geographical facts of existence. To 'begin with the subjective' means, as it were, to start with the objective fact of at least one's own existence. Similarly, existential freedom is not 'free will', or caprice, but responsibility. It means that 'man is condemned to be free', for 'we have no excuses behind us, nor justification before us. We are alone, with no excuses' (Sartre, 1946; Novack, 1966).

The sense of dilemma conveyed by the phrase 'we are alone, with no excuses' perhaps best typifies the mood of existential philosophy and it was, no doubt, that mood which made existentialism a popular post-World War II literary and intellectual movement. By emphasising the culpability of each individual in a world without excuses, and by recognising that condition as one fraught with anxiety and even despair, existentialism championed the individual and his or her freedom against all the external totalitarian forces of the state, of society, and even of nature (i.e., via the aggressive generalising rationalism of the sciences). For this reason it gained a reputation for extremism and, for that matter, fostered a concern for the individual human being in extreme situations where, despite constraints, the individual human consciousness (freedom) prevailed.

Some, especially those on the political left, found this defence of the individual and, curiously, even the defence of the rebel as the product and representation of a bourgeois mentality aimed to reject the progressive consciousness of socialism. Others, especially those on the political right, found the attack on rationalism and the concern for the extreme situation an example of continental, European excess, a return to romanticism, or simply a psychological adjustment to the devastation of European society in the wake of the Third Reich. In either case a fascinating debate emerged around many of the most fundamental principles of Western thought, philosophy and ethics.

One could, of course, pursue that debate here in order to reveal the various dimensions of the existentialist argument. But, in doing so, we would still be left with at least one outstanding question: how and why does that argument impinge upon or relate to the work of geographers? The answer is open to debate, if only because existentialist writers so seldom addressed issues commonly labelled geographical. The identification of geography — the discipline — with the scientific tradition, no doubt itself suggests that existentialism can contribute little to geographical method or theory. Similarly, the concern of existentialists for the 'modern predicament of man', and the strong historical tone of their writings often suggests a type of epochal consciousness devoid of specific geographical content.

Nevertheless, there are several areas of mutual concern where modern geographical theory and existentialism converge to produce both an existential geography and a geographical existentialism. Not the least of these areas, indeed perhaps the most fundamental, has to do with the idea and reality of existential space. And it is toward that spatial dimension of the existentialist argument which we need turn first in order to discern the relevance of any existential geography.

Existential Space

I have argued elsewhere that existentialism begins with a spatial ontology of man (Samuels, 1978). For the sake of clarity we can quickly reproduce the basic elements of that argument here. That spatial ontology was perhaps best phrased by Martin Buber's paper on 'Distance and Relation' (Buber, 1957). Spatiality, he argued, is 'the first principle of human life' and entails a twofold process which he identified as (1) 'the primal setting at a distance' in order to (2) 'enter into relations'. It is, Buber maintained, 'the peculiarity of human life that here and here alone a being has arisen from the whole endowed and entitled to detach the whole from himself as a world and to make it opposite to himself . . .' (Buber, 1957). That initial ability, 'the primal setting at a distance', constitutes the ontological ground of any human existence, for the *sine qua non* of humanness is objectivity, i.e., detachment or estrangement. As Buber argued, however, such detachment does not alone suffice to explain spatiality. Rather, detachment has a purpose, which is to say the second part of spatiality, 'an entering into relations'. Relationship is not conceived here as the opposite of distance, but rather as the goal of the 'primal setting' apart.

In this view, spatiality is more than a necessary condition of human consciousness. It is human consciousness. Spatiality is meaningful in these terms precisely because it constitutes a minimum definition of man as the only historic life form to emerge with a capacity for

detachment. For this reason too the human situation is defined
existentially as one predicated on distance, and estrangement is
understood to be the human situation *par excellence* (Buber, 1957,
pp.97-9; Heinemann, 1958, pp.1-13). At the same time, however, such
estrangement is understood only as the minimum situation, and it is the
goal of existential elucidation to clarify the potential implicit in that
situation, namely relation.

As Buber and other existential writers insist, relationship does not
always or necessarily follow from distance. On the contrary,

> man can set at a distance without coming into real relations with
> what has been set at a distance. He can fill the act of setting at
> a distance with the *will to relation*, relation having been made possible
> only by that act . . . but the two movements can also contend with
> one another, each seeing in the other the obstacle to its own
> realization (Buber, 1957, p.100).

Logically, that is, distance is the precondition of relationship, but the
obverse does not follow. Relation is then not a logical consequence of the
act of setting at a distance, but is rather an existential or even ethical
necessity made possible only by man's will to relate. In Buber's terms,
furthermore, that will to relate is necessary because it is the means
whereby man confirms his own existence in the world.

To put the matter more directly, an animal *needs* no confirmation –
no logical justification – to proclaim either itself or existence *per se*,
for its mental life is one of pure subjectivity or total involvement in its
environment (Buber, 1957, p.98; Scheler, 1970, pp.43-8). Only man
requires confirmation, for only he begins in the world by defining the
environment as opposite to and separate from himself. The need to
confirm arises because without such confirmation, there could not
even be any assurance that there was someone detached, let alone any
assurance that he who detaches himself has any relationship whatsoever
to the world. In Buber's terms, confirmation comes precisely to the
extent that man exercises his will to relate or, in other words,
*endeavours to mitigate distance through relationships with his
environment.*

Though not often emphasised in critiques of existentialism, the
'will to relate' is the core of an existential ethic. Summarised by
Martin Heidegger (Tymieniecka, 1962, p.62), that ethic states that
'the world itself in light of its intentional reference (the subject in
relation) is understood not as a sum of objects or matter – a brute

physical reality – but as a net of relations between man . . . and the realities of his surroundings, as objects of his concern (Tymieniecka, 1962, p.62). Man, in short, either chooses to relate or does not, and the only existential compulsion is concern. Concern is the energising force behind man's relationships and the means whereby detachment (distance) is either overcome or mitigated. In existential terms, such are the rudiments of spatiality, which is to say the dynamics of a human consciousness.

As we have seen, distance and relation, the two components of space, are not simply quantitative designations, but are rather loaded with human or qualitative content. Space is burdened with ethical, spiritual and emotional, as well as rational, significance, for spatiality lies at the root of the elemental human dilemma of estrangement. Staying with Buber's phraseology, if that dilemma is resolvable, it finds resolution insofar as the two components of spatiality – distance and relation – are dialectical concomitants of one another. What distance necessitates (detachment), relation fulfills (belonging) so that 'distance provides the human situation, (while) relation provides man's becoming in that situation' (Buber, 1957, p.100). In consequence, (1) there is by definition no such thing as pure subjectivity (relation without distance) in a human consciousness, but (2) pure objectivity (distance without relation) is either meaningless or contrary to human history. Man is ontologically the spatial being *par excellence*, because he is existentially tied to the encounter with distance. Similarly, insofar as the phenomenon 'space' is human in origin or propagation, so too is spatiality always a reflection of the dialectic of distance and relation. For this reason the meaning of space is 'existential', which is to say a function of the human encounter with distance and its fulfilment in relationship.

As Sartre restated the existential ontology, 'human reality is the being which causes place to come to objects' (Sartre, 1966, p.370). What this means is that 'to come into existence . . . is to unfold my distances from things and thereby cause things "to be there"' (Sartre, 1966, p.407). To unfold distance is the *sine qua non* of a human consciousness, but in so doing man engenders at least two places, i.e., the place (hence existence) of the object, as well as the place (hence existence) of the one who emplaces. The latter is 'subjective' in the sense that it entails 'an engaged upsurge in a determined point of view' (a place from which consciousness emerges), and insofar as place is always an assignment by someone (Sartre, 1966, pp.307 and 407). As Sartre added, 'the place of an object or instrument, even if sometimes precisely assigned (i.e., through agreement with others) does not derive from the nature of the object itself, but it is through me that place . . . is realized' (Sartre, 1966, p.370). As such, emplacement is always a reference to something from someone, and that reference is the link between object and subject, distance and relation. Most generically, the concatenation of such reference points is the organising principle of 'existential space'.

Just as there are two components of spatiality (distance and relation), so too are there two aspects of existential space. The first of these emphasises assignment and can be described as 'partial space', whereas the second emphasises situational contingencies and can be described as 'situations of reference'. For the sake of brevity they are perhaps most simply defined as follows:

(1) *partial space* is that net of relationships between man and the world where the latter is an object of his concern. It is 'partial' both in the sense of bias or subjectivity, and incompleteness.
(2) *situations of reference* are the historic conditions within which assignments arise. Our partial spaces are rooted in and references to situations which are not solely dependent on us, but which we make our own through relationships of concern. Without such roots we have nothing with which to relate or about which to be concerned.

Though they employ other terms, Merleau-Ponty, Sartre and Jaspers each discuss problems associated with partial space and situations of reference in great detail. The former, as Merleau-Ponty would have it, is distinct from 'clear space, that impartial space in which all objects are equally important and enjoy the same right to existence'. (Merleau-Ponty, 1962, p.287). Rather, it constitutes a second space which 'cuts across visible space [and] which is ceaselessly composed by our way of projecting the world' (Merleau-Ponty, 1962, p.287). In terms of spatial perception, partial space is accomplished by focusing or noticing, which is to say by ordering the world. Here our partiality intervenes to prevent chaos, for a world in which all objects are equally important is anarchical. Effectively, partial space is the assignment of meanings to places. Some places (assignments) are more meaningful than others.

Such assignments do not represent an indulgence in pure subjectivity, but rather represent a reference to something or someone already other than and at a distance from oneself. Partial space, in short, is contingent on references at a distance. Such references are composed of situations into which we are thrust, over which we may have little or no control, and about which we may be imperfectly aware. We are, as it were, born into a world which only later becomes our world. The implications, as Jaspers warned, is that 'my place is . . . determined by coordinates; what I am is a function of this place; existence is integral and I myself am but a modification or a consequence or a link in the chain' (Jaspers, 1957, p.27). Though it is always 'my' place, this place has its foundations in a world apart from 'me'.

There is another way of saying this in terms of a human consciousness. The latter is not infinite in the sense of timelessness or spacelessness. Rather, it has an historical beginning and end, and our way of projecting the world is tied to this fact of historicity. Space – the fact of consciousness – is inextricably linked to time – the emergence of consciousness. As Jaspers would have it, what this means is that our partial spaces have their roots in historic situations (Jaspers, 1969, vol.2, pp.104-29 and pp.342-59). Existential space, as such, is always a reference to the world which becomes our world only after the fact.

Comprised of meaningful places, existential space must nonetheless be understood in terms of historic contingencies or, in other words, references to situations which only become ours. Composed of these situations, however, existential space is always also reflective of someone's partiality or the meanings people give to places. Home, the 'first place', is, for example, the situation of our beginnings, but it is also the first 'orientation' toward a world which we make our own (Bachelard, 1969, pp.3-37). Places not (or not yet) our own follow suit at whatever scale. In this fashion, the physicist's 'infinite universe' is infinite only to the extent that the universe is made up of unlimited reference points with which to relate. The theologian's infinite God is similarly not 'nowhere' beyond space and time, but 'everywhere' in space and time available for relationship.

Unlike the Kantian segregation of space and time, subject and object, the existential dialectic maintains that these are in fact concomitants of one another. Time is always some place and place is always some time. Moreover, emplacement (the partial or subjective act of assignment) is here always a reflection of historicity. The resultant 'geography' is at once subjective as a reflection of partiality (an ensemble of presumably meaningful reference points), and objective as a reflection of the historical conditions into which some consciousness (if only that of the geographer) has been thrust. The careful review of virtually all introductory geography texts will suffice to offer any number of

examples of 'existential space' simply by virtue of all the places left
unmentioned. Similarly, the use of place names or regional
designations (e.g., the Free World, the Third World, etc.) and the
space or time allocated such places will reveal historical co-ordinates
relevant at least to the author of such geographies.

All spatial relations, arrangements and attachments can be
interpreted in the light of the idea of existential space. There is little
extraordinary, mystical or romantic about the latter, for at root
existential space is nothing other than the assignment of place. Who
makes such assignments arise can often seem exceptionally difficult
to discern. The means to that discernment and especially its utility
in geography can, nonetheless, be discussed in terms of an
existential method that results in what I have elsewhere described
below as 'the biography of landscape' (Samuels, 1979).

An Existential Method

By beginning with the subjective and by placing existence before
essence, the existential argument rejects any sentiment that would
abandon man to nature, or nature to some abstract logos lacking in
human signification. It thereby emphasises a type of perspective
aimed to defend anthropocentrism. Furthermore, by introducing a
spatial ontology of man and a human definition of spatiality, the
existential argument serves to underscore any system of thought
poised to articulate the variegated meaningfulness of spatial
arrangements, relations and attachments. As such, existentialism
provides a firm foundation for a philosophy of human geography.

What remains, however, is clarification of the methodological
implications of any such existential philosophy of geography. The
operative principle of any existential method is that man
'expresses himself as a whole in even his most insignificant and his
most superficial behavior. In other words, there is not a taste, a
mannerism, or a human act which is not revealing' (Sartre, 1966,
p 726). Consciousness and behaviour, that is, are always revealing
of the way man distances and enters into relations with the world.
Spatiality, the act of assigning places, and behaviour in space are
always here 'an expression of the total life of the subject (Merleau-
Ponty, 1962, p.283; Sartre, 1966, p.726). They are reflective of the
way in which the subject organises the world as an object of concern.
An existential method is one that endeavours to 'decipher' that total

expression, and to do so by 'beginning with the subjective'.

To 'begin with the subjective', once again, means that 'man first of all exists, encounters himself, surges up in the world – and defines himself afterwards' (Sartre, 1946, p.290). In the process '. . . man's inward attitudes, the way in which he contemplates his world and grows aware of it, the essential value of his satisfactions are the origins of what he does' (Jaspers, 1957, pp.4 and 175). This does not mean, however, that the world itself has no 'objective' content. Rather, it means only that the world has no objective centre – that space and time are configurations of reference from someone to something. In the words of Karl Jaspers:

> The world now without an objective center centers everywhere;
> and I once more in the middle of it, though no longer objective
> in a sense that applies identically to everyone. The only center
> is the one I occupy as an existing individual. My situation is
> what I start from and what I return to, because nothing else
> is real and present; but the situation itself becomes clear to me
> only when I think with reference to the objective being of
> the world . . . I can neither grasp my situation without
> proceeding to conceive the world nor grasp the world without
> a constant return to my situation, the only testing ground for
> the reality of my thoughts (Jaspers, 1969, vol.1, p.106).

An existential method, in short, begins by examining the centres people (and, in particular, individuals) occupy and the way they define their relations with the world. It begins with an analysis of existential space, by first analysing the partiality that people project into their situations. More specifically, the method entails what Sartre called 'existential psychoanalysis' (Sartre, 1966, pp.712-34).

We need not explore all avenues of 'existential psychoanalysis' to trace some of the more salient points for an existential method in human geography. At the outset, the method begins by reconstructing the partial spaces of individuals in order to discover there the roots and goals of relationship. At this point, the existential method veers sharply away from scientism, and from various schools of modern psychology. As Sartre argued, where psychology becomes 'an allusion to the great explanatory variables of our epoch – heredity, education, environment, physiological constitution', or where it culminates only in 'a list of behavioral patterns, drives and inclinations', it fails to give us an understanding of the concrete, historical, centre of consciousness

(Sartre, 1966, p.715). The first task of an existential method is 'not to establish empirical laws of succession, nor [to] constitute a universal symbolism', but to rediscover at each step a symbol functioning in the particular case which some subject is considering (Sartre, 1966, p.732).

The existential method thus emphasises individuality. This does not mean, however, that it deals only with individuals. Rather, as Jaspers argued, by means of 'communication' individuals share their concerns and reference points, and thereby enter into relations with one another (Jaspers, 1969, vol.2, pp.47-103). Such communication gives rise to groups, but the latter are never simply aggregates. They constitute groups precisely to the extent that the people have common concerns and reference points. For this reason, too, as Sartre explained, the existential method can be broadly comparative, so long as that comparison does not end in abstract archetypes that sublimate individuals and concrete (specific) groups (Sartre, 1966, p.727). To 'begin with the subjective' here means that, as we discover symbols that function in particular cases, we will find that the meaning of those symbols sometimes converge toward particular groups, and sometimes diverge toward specific individuals. The point is, however, 'if each being is a totality, it is not conceivable that there can exist an elementary symbolic relationship . . . which preserves a constant meaning in all cases; that is, which remain unaltered when they pass from one meaningful ensemble to another ensemble' (Sartre, 1966, p.732). No all-inclusive iconography will suffice to explain the meanings people give their reference points.

The meanings people give their reference points vary greatly from individual to individual and from specific group to specific group. Where constancy of meaning is detectable, it is indicative of some individual's or some group's commitment to a particular set of reference points and relationships. Where those reference points change, as in the event of war, migration, travel, or even mood, the symbolic content of existential space also changes. Hence, for example, few North Americans may ever have imagined a place called Vietnam until, after having already crept into their situation, it became a daily event colouring their existential spaces. The 'situation' Vietnam was there, building its way toward an intersection with the lives of countless strangers, to become a meaningful reference point. Once the American 'involvement' began to end, however, the symbolic content of 'Vietnam' shifted away from images of falling dominoes, war, death, national liberation, and so on. It became instead an aspect or a reflection of an American self-consciousness identified with participatory democracy, student rebellion, counter culture movement and so on. The place Vietnam, that is, left the Asian continent to arrive squarely in America.

There are, no doubt, a number of apparent similarities between the existential method described here and the behavioural analysis of personal and social space. Agreement exists, for example, with Robert Sommer (1969) and Claude Levi-Strauss (1963) on the need to discover the symbolic content of spatial arrangements. The view that it is necessary to decipher the meanings people give their places finds similar concurrence with those interested in spatial and environmental perception (Hall, 1969; Gould and White, 1974). Cognitive mapping, for example, would seem to be an appropriate technique associated with the existential method.

For all such appearances, however, behavioural analysis is not coincident with the existential method. Indeed, there are considerable differences and at least one fundamental disagreement between the two. The latter can perhaps best be illustrated by reference to one of the more recent methodological and substantive statements in behavioural geography, notably David Harvey's *Social Justice and the City* (1973).

Harvey's conclusion that '. . . space only takes on meaning in terms of significant relationships and a significant relationship cannot be determined independent of the cognitive state of some individual and the context in which that individual finds himself' (Harvey, 1973, p.34), reads remarkably close to our definition of existential space. Similarly, his statement that 'the city contains all manner of signals and symbols' and 'we can try to understand the meaning people give to them', is virtually a restatement of what has been described here as an existential method (Harvey, 1973, p.32).

Once such statements are carefully probed, however, they reveal a sentiment quite unlike that identified with existentialism. In Harvey's case, his is a concern focused primarily on the way in which spatial symbols affect behaviour and with 'the message people receive from their constructed environment' (Harvey, 1973, p.37). Most critically, the source of these symbols and meanings is itself cast away in the single assertion that he (Harvey) doubts 'very much whether we will every truly understand the intuitions which lead a creative artist to mould space to convey a message' (Harvey, 1973, p.32).

The assumption, endemic to Harvey's perspective, that the source of creativity is 'intuitive' and thereby inaccessible, is wholly unacceptable to the existential method. It necessarily leads Harvey to the conclusion that we need only find 'techniques to measure the impact of spatial and environment symbolism' (Harvey, 1973, p.32). It leads him further to suggest that, 'while it would be best to evaluate the cognitive states of the individual with respect to his spatial environment . . . practical considerations make it impossible to use anything other than overt behaviour when large aggregates of population are involved' (Harvey, 1973, p.33). In each case, in short, man is viewed as the result of some external stimuli (now termed 'spatial symbolism'). Furthermore, only a general methodology that sublimates individuality to some universal syntax and to the expedience or sufficiency of data (aggregates) will suffice to explain human behaviour in space. In Buber's terms, the only space measured here is the space of distance.

If, however, we turn Harvey's methodology back on its feet grounded in the concrete experiences of individuals and specific groups, we will find that his already given city filled with signals and symbols was, in fact, given by architects, planners, builders, dreamers and doers of various deeds. Their so-called 'intuitions' are hardly inaccessible to those willing to look. More often than not, the meanings they intended are more accessible than the mystical intuitions of mass man, especially as they leave a record of articulated rationalisations. Hence, for example, we need not guess the symbolic intent of freeways and parks in the city of New York. The voluminous personal and public archives of a certain Mr Robert Moses, who was responsible for much of that landscape, are clearly available (Caro, 1975). An analysis of his intentions, as well as the meanings other people give the result of his efforts, is hardly impossible. It requires only a sensitivity to the *biography of landscape*, which is to say the existential origins of spatial arrangements, relations and attachments.

The Biography of Landscape

Every ensemble of meaningful places, together with the situations from which they emerge, here constitutes a landscape with a biography. The analysis of that biography may be either backward from an already given landscape to particular individuals or groups, or forward from the latter toward landscape articulation. In both cases, the concern is to identify the sources of a landscape and the meanings that landscape conveys.

To the extent that existential space operates at any human scale, we can look to individuals or specific groups (large or small) for the biographies of landscapes. Though we are constrained to be consistent in the scale of analysis, we can investigate the 'great' figures and civilisations or any number of 'lesser' individuals and communities. Every landscape is someone's existential space. At the same time, however, everyone's landscape is a reference to a geography already given. The landscape of modern China, for example, can be investigated in the light of the intentions and meanings given it by Mao Tse-Tung. Unfortunately, the notes he used when he taught a course on the geography of China are not yet available, but other sources that reveal his views on the matter are accessible in the extreme (Tse-Tung, 1965, vols.1-4; Murphey, 1967). In addition, there is a wealth of data on the impact of those views. But Mao Tse-Tung's views did not emerge in a vacuum. As Howard Boorman suggested, they represent 'the distinctive product of an interlude of massive disorientation in Chinese life marked by the breakdown of Confucian patterns and by the search for new political, intellectual, and ethical reintegration' (Boorman, 1968, p.307). The places of that reintegration — Peking, Shanghai, Canton, Chao-shan, Yenan and so on — were there and are still there, but now transformed in the vision of Mao Tse-Tung and those who shaped and followed him.

The biography of the landscape of contemporary China may then have its beginnings in an analysis of the man who Westernised China by Sinifying Marxism. It does not, however, end there, for the roots of that biography are deeply embedded in the history of China — in the Confucian patterns that broke down at the end of the nineteenth century, as well as the Legalist and Confucian patterns that rationalised and helped give rise to the imperial-bureaucratic order in early Chinese history. Here too that biography can be traced to particular individuals and groups — to Confucius, Hsun-tzu, Meng-tzu and many others. Furthermore, that biography also has its roots in the history of those, like Augustine Heard, David Sassoon and other China Traders, who helped build such great cities as Shanghai and T'ientsin. There are, in effect, as many landscapes of China as there are those who related to China.

To be sure, some of these landscapes are more 'relevant' or 'important' than others, especially insofar as the individuals and groups concerned can be said to have had an impact on the landscape of China. But the point remains the same irrespective of such impact. Every landscape of China is the distinctive product of some perspective on Chinese history, and the history of those who have either influenced or been influenced by China. In this sense, everyman's China (the China of *Time* magazine) is no less 'real' than the Sinologist's China (the China of the *Journal of the American Oriental Society*). The latter is perhaps more 'correct', but primarily because, for Sinologists, China represents a consistent and most meaningful reference point. They can be said to 'know' more about China than the 'layman', because they are committed to that reference point.

The landscape such 'knowledge' intends may be objective, in the sense of knowledge shared among many, or subjective, in the sense of knowledge peculiar to one person. The meaning of the New England and Missouri landscapes, for example, may differ as between Mark Twain and T.S. Eliot. In Eliot's case, as well as Twain's the Mississippi River left an indelible impression, but for Eliot it was an urban image, rather than Twain's rural Missouri. As Eliot put it,

> My urban imagery was that of St. Louis, upon which that of Paris and London have been superimposed. It was also, however, the Mississippi as it passes between St. Louis and East St. Louis in Illinois: the Mississippi was the most powerful feature of Nature in that environment. My country landscape, on the other hand, is that of New England, of coastal New England, and New England from June to October. (Eliot, 1960, pp.421-2).

Later, 'English landscape [came] to be as significant for me and as emotionally charged as New England landscape' (Eliot, 1960, pp.421-2). The meaning of these landscapes for Eliot is also clear, for as he explained his poem 'The Dry Salvages', the poem 'begins where I began, with the Mississippi; and it ends, where I and my wife expect to end, at the parish church of a tiny village in Somerset' (Eliot, 1960, pp.421-2).

The biography of a landscape may, in this sense, tell us as much about its author, as about the place itself. In cases like that of T. S. Eliot this only seems appropriate, for our attentions are drawn more to him, the poet, than to either Missouri or New England. But the principle of authorship holds for all such landscapes, even when the author is less than famous. The much used cognitive map of the New Yorker's view of the United States, for example, not only tells us something about both New York and the US, but also about a parochial view of New Yorkers (Willingform, 1936, reprinted in Gould and White, 1974, p.38). In a more complicated fashion, every 'geography' reveals the often hidden reference points meaningful to its geographer. Just as maps entail the bias of projection, so too are 'geographies' made in the wake of 'an upsurge in a determined point of view'. How many 'urban geographies' are written by those who have yet to live in a 'city'?

How many more are written from the vantage of one or two particular cities? How many others have been written with an anti-urban bias or with a bias predicated on the historical circumstances of large North American cities in the 1960s? How many 'geographies' have been written from the perspective of 'norms of behaviour' on the part of middle-class America? We could go on, but the point would remain the same. All such 'geographies' have a biography, and elucidation of that biography is what is intended by the existential method.

An Existential Geography

The discussion so far, albeit brief and oversimplified, has already revealed some of the ways in which the existential argument intersects with and provides a philosophy and method appropriate to geography. By way of summary, it may help here to emphasise the point that ultimately an 'existential geography' is a study in the biography of landscape. The two most important aspects of that geography are (1) it begins with the subjective or with the issue of authorship in order (2) to discover the relations individuals and specific groups have with their environments as objects of their concern. A biography of the former is here always a history of the latter. Effectively, *an 'existential geography' is a type of historical geography that endeavours to reconstruct a landscape in the eyes of its occupants, users, explorers and students in the light of historical situations that condition, modify, or change relationships.*

If the biography of some landscape is here necessarily particularistic, this does not mean that the spatial arrangements, relations, or attachments in the landscape have no general validity. On the contrary, they acquire general validity at least insofar as the historical situations that set the context of their rise and concatenation has some general applicability. In principle, every landscape biography acquires some general validity in the light of shared historical contexts. There is a geography of North America, for example, predicated on the experiences of a generation moulded in the context of the Great Depression, in the same sense that there is a post-World War II geography of North America that has its roots in the experience of affluence. In this fashion, the geography of North America is a geography made in the wake of these two orientations, and their ramifications are manifold.

Such shared contexts in a world predicated on mass communication may be expected to have even the most general of ramifications for

/

an existential geography. Increasingly, with modernity reaching even into the most secluded parts of earth space, all the certainties of a landscape formed under the tutelage of what Paul Wheatley and René Berthelot have described as 'astrobiological' orientations focused on the sacred, have collapsed (Wheatley, 1976; Berthelot, 1949). At the same time, to paraphrase C. Wright Mills, 'the sanctions or justifications for the new routines (man) lives and must life', either remain in ferment or simply fail to take hold (Mills, 1951, p. xvi). The landscape, in the meantime, remains unfixed or caught *in extremis* between poles of differing orientations. That situation, commonly referred to in existential literature as modern alienation, also has its geography, the map of which is filled with dislocation (the lack of place or the misappropriation of place). For some, that map has no meaningful reference points or points that bind, and the geography of modernity is a 'wasteland', the contours of which are formed by all the dead and dying reference points. For others, the map conveys all the human endeavours to find new reference points in the midst of changing historical contingencies. If either map has any general validity, however, it is because the mediating force – history – has impinged on particular individuals, groups, communities and peoples in many places, though often in different ways and at different rates.

General applicability is then not lost to an existential geography. Nevertheless, it is important to keep in mind the fact that generalisation *per se* has no special utility in an existential method. Indeed, as with landscape biography, even the emphasis on shared contexts reconfirms a particularistic (regional) approach. Missouri and New England, for example, are the regional contexts of T. S. Eliot's landscapes, whereas California and New England serve as contexts for Robert Frost. Eliot and Frost may share New England, but then again, perhaps because they do not share Missouri and California, their respective 'New Englands' are somehow different. For much the same reason, even if we shift the scale of our concerns toward such global issues as the historical contexts of modernity, the regional implications are nonetheless clear. A geography of modernity predicated, for example, on the experience of the Great Depression would be a regional geography confined to the people and places affected by the economic collapse of many Western economies. The experience of deprivation, however, was not uniform even within the centres of commerce, let alone areas then hardly touched by international finance. Similarly, the first and second 'world wars' were not global events, but rather constituted 'civil wars' fought largely in the context of Western civilisation or, as in the Pacific, between established Western authority and an emergent Westernising force. Here, too, the experience was not felt uniformly throughout. For some, a geography of the last European civil war might have its centre in the London Blitz, in a Nuremberg beer hall, in the factories of America, in Stalingrad, or on the front lines themselves. For others, the centre shifted in the grotesque landscape of the Third Reich to a Dachau, a Treblinka or any number of other such demented cores. Though linked by the rise of the Nazi Party, the existential geographies of the Second World War in Europe are thus manifold and relative, which is to say particular and regional.

There are, no doubt, various and monumental dangers associated with the insistence on such relativity of context and partiality. We may recall that an Adolf Hitler, his chief political geographer, Karl Haushofer, and the grand architect of the Third Reich, Albert Speer, endeavoured to design the landscape of the world according to the principles of partial space and situational reference. The history of Germany and of Europe as a whole allowed the choice of any number of other reference points, but these individuals and the Nazi ethos as a whole chose to fill the landscape with a particular set of meanings. The existential method insists here not on the type of choices made, but on the fact of choice itself. It makes clear what history — in this case, the history of modern Europe — already demands. For good and evil, people — individuals and groups — whether sane or insane, rational or irrational, well-intentioned or demonic, make their choices and their landscapes. The result, simply and most emphatically, is that for every landscape or every existential geography there is a someone who can be held accountable.

Through the examination of the biography of landscape an existential geography demands not so much the elucidation of 'cause', as the assignment of responsibility. Ultimately, what 'causes' an individual or a group to construct its landscape and to fill it with meanings, though partly discoverable by means of the method described here, has no all-encompassing utility. The events which give rise to those 'causes' are not, that is, totally repeatable in every case. Furthermore, unless one is convinced of the perfectability of man so that, for example, the discovery of uniform causes may provide the means to change such 'causes', the issue is itself moot. Whether 'cause' can be changed or manipulated, individuals and groups change, and as they change so goes the assignment of responsibility. In existential terms, that is all and that is enough.

Conclusion

There are many other dimensions of the existential argument appropriate to consideration by geographers. Though they deserve attention, they would not be discussed in this chapter. Hopefully, the broad outlines of both the philosophy and method of an existential geography as discussed here will encourage a consideration of the strengths and weaknesses of that geography. In the meantime, it may suffice to emphasise one point. The existential philosophy and method is both objective and subjective, historically focused and fundamentally concerned with the nature of spatiality, but most of all it is anthropocentric. Through the encounter with the spatial ontology of man, his partial spaces and situations of reference, an existential geography intends most to emphasise the human core of existence. If there is one singular advantage of such an emphasis in geography, it will be to expand the horizon of geographical research to encompass understanding and faith, rationality and irrationality, knowledge and feeling – the whole spectrum of human emotions and capacities as subjects of concern. In the end, we may after all break through to what Carl Sauer once described as the 'art of geography' which would further 'lead to philosophical speculation'. We can, in conclusion, ask the same question he posed: 'And why not' (Sauer, 1956)?

IDEALISM

Leonard Guelke

Introduction

Over the past two decades the discipline of geography has undergone
many changes associated with the adoption of new techniques and ideas.
The 'quantitative revolution', theoretical geography, behaviourism,
Marxism and phenomenology are some of the movements and themes
that have transformed the modern discipline. The discipline has
changed from a relatively cohesive field into an incredibly diverse
collection of related and unrelated topical studies held together, if
at all, by a common concern with spatial relationships. With the
proliferation of new approaches and methods, the traditional concern
of the geographer with the earth as the home of man, with man/land
relationships and with landscapes and regions has been largely neglected.
An important reason for the decline of interest in the traditional core
area of the discipline has been the lack of an appropriate philosophy
and methodology for such studies. A methodology based on the
philosophy of idealism, I believe, could provide the foundation of a
revitalised, 'traditional' geography which is both academically coherent
and intellectually stimulating.

The Idealist Philosophy

In a philosophical sense idealism is the view that the activity of mind
is the foundation of human existence and knowledge (Acton, 1967,
p.110). Idealism comprises two distinct philosophical positions known
as metaphysical and epistemological idealism. The metaphysical idealist
maintains, in opposition to the proponents of naturalism and
materialism, that mental activity has a life of its own which is not
controlled by material things and processes. The epistemological
idealist holds that the world can only be known indirectly through
ideas. On this view all knowledge is ultimately based on an individual's
subjective experience of the world, and comprises mental constructs
and ideas. There is no 'real' world that can be known independently
of mind. An acceptance of epistemological idealism does not entail
the acceptance of metaphysical idealism, or vice versa (Acton, 1967,

p.110). The foundation of the approach to human geography advocated here is epistemological idealism.

The idea that human behaviour is largely controlled by mental activity is the basis on which idealists insist that the social sciences and history are logically separate from the natural sciences. The logical positivist idea of a single unified science is rejected in favour of an autonomous social science (including history) with its own approach and methods. Although human behaviour cannot be treated as a material process in the normal (natural) scientific way, the rational character of human thought makes it possible for one to understand deliberate human activity in a way that it is not possible to comprehend material processes. A number of idealist philosophers have developed distinctive methodologies for the social sciences and history on the assumption that human activity must be understood in terms of thought. The idealist position does not involve a rejection of the broadly scientific goals of true explanations supported by appropriate evidence, but rather it maintains that these goals can only be approached in human studies by adopting methods that take the special nature of thought-controlled, human behaviour into account.

The idea that the social sciences (*Kulturwissenschaften* or *Geisteswissenschaften*) are logically independent of the natural sciences is not new. In the late nineteenth and early twentieth centuries this position was expounded in considerable depth by neo-Kantian philosophers, notably Dilthey and Rickert, and by the Italian philosopher Croce (Croce, 1941; Gardiner, 1959; Rickman, 1961; Rickert, 1962). This work with that of others, was developed and extended by the English philosopher R. G. Collingwood. The position I expound here is largely derived from the writing of Collingwood, especially his posthumously published work *The Idea of History* (1946). In developing an idealist approach my major concern is with methodological questions, that is, with questions that relate to the actual application of the approach to concrete situations.

The Method Rethinking or Verstehen

The idealist maintains a human action or its product is understood by a scholar rethinking or reconstructing the thought contained in it (Collingwood, 1956, pp.213-15). This mode of explanation is often referred to as *Verstehen*, understanding, and is contrasted with the theoretical and nomothetic approach of the natural sciences. In using the method of *Verstehen* a researcher does not have *carte blanche* to impute any thoughts he chooses to his subjects. The human geographer or social scientist is concerned to discover 'what really happened' and his interpretations of the thoughts of others should be taken no further than the evidence sanctions. One can look upon the search for a rational explanation of a specific action as being akin to the natural scientist's search for a theoretical explanation of a natural phenomenon. In both cases an imaginative effort is involved to find an explanation that fits the facts. In seeking such an explanation in terms of thought, the investigator needs to conduct a critical dialogue with his evidence, and the inferences he makes from it about thought need to be subjected to critical scrutiny and cross-examination.

The idealist maintains that it is possible to reconstruct and understand a logical sequence of thought of another person in a way that it is not possible, for instance, to re-experience his emotional life. Thus, one can understand why Columbus sought to reach China by sailing westward across the Atlantic Ocean; it was a rational act of an individual who believed the world was round but thought it was considerably smaller than its actual size. In this case, understanding is achieved by rethinking the thought pertinent to the problem Columbus faced and following the logical development of his ideas. The fact that the idealist restricts himself to thought that is in some way rational is not the severe limitation it might at first sight appear, particularly in human geography. Geographers are concerned overwhelmingly with activities which are the outcome of deliberate rational actions. When people grow crops, build houses, move to new homes and exploit resources their actions are the result of rational thought and are therefore, in principle, open to being understood in terms of the idealist method rethinking.

An objection to the concept of rethinking or *Verstehen* is that it ignores emotional and psychological aspects of human behaviour. This charge is correct if it is meant to suggest that the idealist cannot rethink human emotions. It is not correct if it is meant to imply that the idealist must entirely neglect any non-rational attributes of humanity. The idealist will generally base his interpretations of thought on the hypothesis that an individual or group possesses 'normal' physical and emotional attributes, but would modify this assumption should there be evidence to justify a change. For example, if medical evidence were produced to show peasant cultivators suffered from a debilitating disease, their lack of interest in agriculture would likely be attributed to physical causes not mental ones. Similarly, it might be necessary to take account of psychological factors in a situation which involved 'abnormal' stress. In these cases the idealist would recognise that physical and emotional factors, as it were, assumed a greater than usual dominance over mental ones and thereby negated the usual assumption of rationality (Guelke, 1974, p.194).

The key concern of an idealist geographer is not with providing a causal explanation of an event (in the sense of listing all the necessary and sufficient conditions for its occurence) but rather in elucidating its human meaning or significance. Such elucidation must be done in relation to thought. For example, the human significance of a volcanic eruption or a nuclear accident is decided by the meanings people attach to these events. The 'real' events (which, in any case, can only be known indirectly once they have happened) are only of importance in so far as they are endowed with human meanings. These meanings will vary in relation to the ideas and backgrounds of those who might be concerned about them. Thus a volcanic eruption might lead people to abandon their homes; a nuclear accident to a change in energy policy. However, the actual reaction of people to such events is not a function of the events themselves, but, rather, of their perception and understanding of them. The task of the human geographer is basically that of elucidating what the world means to its peoples and showing how human activity on the land is related to the way people have construed their situations. Such elucidation is achieved by a geographer rethinking the thoughts of those whose actions he is investigating.

The essence of rethinking thoughts involves discerning the human purpose or rationality of specific acts; it does not involve the impossible feat of recreating all someone else's actual thoughts in one's own mind. This point is nicely illustrated with an example from Collingwood's own work on Roman Britain (Collingwood and Myers, 1937, pp.124-34; Goldstein, 1970, pp.27-31). The example concerns Collingwood's attempt to find an explanation for the existence of a continuous earthwork (the Vallum) that runs parallel to Hadrian's wall on its southern side. Collingwood, having rejected the notion that the Vallum could have had any defensive value, argued that it was probably created to avoid any confusion arising between the military and financial authorities in carrying out their respective frontier responsibilities. The wall itself, Collingwood maintained, might indeed have served both functions, but such an arrangement would have left doubt as to the undisputed authority of the commandant in military matters. So Hadrian had an additional barrier built at which financial matters of the frontier administration could be transacted. Collingwood noted, however, that there is no proof that his explanation is correct, 'but that it fits the facts' (Collingwood and Myers, 1937, pp.124-34; Goldstein, 1970, pp.27-31).

In the above example Collingwood has sought to provide a rational explanation of a human creation in terms of the probable purpose it was intended to serve. The essence of Collingwood's approach is well summarised by Goldstein:

> If Collingwood's solution to the Vallum problem is correct, there is a clear sense in which he has rethought the thoughts of Hadrian in all of their historically-relevant character. The essential considerations which presumably passed through Hadrian's mind as he came to the decision to have the Vallum built have passed through Collingwood's as well. There is no suggestion of his having entered fully into the existential experience of the historical actor in the sense of reproducing the feelings, emotions, and other appurtenances of existing-and-experiencing-here-and-now in the way that an historical novelist might seek to do. Here, it seems, is the central feature of what Collingwood thinks history is: it is rethinking thought on the basis of evidence without ever becoming psychology. As rethinking thought, its object is what can be detached from the original context of action and be reproduced in the later context of historical inquiry, hence its object is universal and not existential (Goldstein, 1970, p.32).

In human geography the explanation of man's activities on the land will often follow the form of the above illustration, which should be examined in its entirety if the full power of Collingwood's method is to be properly appreciated (Collingwood and Myers, 1937, pp.124-34; Goldstein, 1970, pp.27-31). Indeed, geographers are often faced with explaining the existence of concrete, man-created objects which are in many ways similar to the case I have just discussed. Different people in making use of the earth have, for example, created distinctive field and settlement patterns. These patterns are not arbitrary, but reflect the thinking of the people who created them. In brief, a man does not make a fence, put up a barn or clear a forest without good reason. It is the geographer's task to find out what those reasons were. In many situations it will involve historical research, because the purpose that, say, a building or road was originally designed to serve might no longer be related to the landscape of the present.

The idealist geographer does not neglect the material conditions of human existence, but insists that such conditions acquire their significance and meaning in terms of human desires and ideas. The functional concept of resource use developed by Zimmermann is a good example of an approach to the material world based on an implicit acceptance of an idealist point of view (Zimmermann, 1964, pp.1-61). This concept of a resource emphasises that a material object can only be considered a resource if it can be made to serve a useful purpose through human ingenuity. For example, the interest that developed in the rubber trees of the Amazon towards the middle and end of the nineteenth century was a direct result of the invention of the vulcanisation process (Zimmermann, 1964, pp.19-20). This rational invention of the human mind rendered useful a part of nature that had previously not been much used in meeting human needs. The extent of the exploitation of rubber in the Amazon was in turn dependent on the demand for this product created by another series of inventions of which the automobile was the most important. A full account of the extent and nature of the exploitation of the Amazon forest would show how contemporary ideas on environment, technology, economics and politics influenced the perception and strategies of the people who were directly or indirectly involved in the extraction of rubber from this area.

The cultural context within which human activity takes place is of crucial importance in idealist explanation. A culture is based on a set of shared ideas and assumptions which form the basis of what one might term a 'world view'. A world view incorporates a large variety of interrelated ideas and shapes the way in which individuals and groups interpret the world around them. Thus, for example, the world views of Russians, Americans or Somalis are all based on distinctive systems of ideas. It is not possible to extract ideas from their context without destroying their essential meaning. Russian ideas about industrial development must be understood within the context of Soviet thought. American ideas about pollution must be seen against a background of their legal system and the protection it affords individual citizens. Somali ideas on land must be understood in terms of their pastoral economy and social system. In brief, the idealist geographer recognises the importance of studying human activity on the earth in relation to the overall cultural context. Although this does not preclude generalisations it does mean that the validity of the generalisations that are made will probably be limited to regions and peoples of broadly similar culture.

A prime goal of the idealist geographer is to elucidate the meaning of human activity in its cultural context. The events and phenomena of the world acquire significance and meaning for individuals and groups in terms of their ideas and theories. Thus an idealist geographer is not interested in why, for example, wheat prices fluctuate or in investigating the causes behind such fluctuations, but rather he is interested in the meaning these fluctuations might have for the peoples he studies. In this case the key questions would be: 'How do the people under study interpret this phenomenon?' and 'What are the behavioural consequences of their interpretation of it?' The same approach would underlie the investigation of all phenomena and events which were thought to have an influence on human activity from volcanic eruptions to failures of nuclear power plants.

A general criticism of the idealist approach is that one can never know for sure whether one has actually succeeded in providing a true explanation. This charge is undoubtedly correct, but on closer examination it loses much of its force as an argument for not taking the approach seriously. Although one can never know with certainty that an idealist explanation is true, the same objection is applicable to all interpretive and theoretical work. The theoretical physicist can never be certain of his theories; he can never be certain that inferences he makes have a solid foundation in the 'real world'. Indeed, the history of natural science is largely a history of abandoned theories. Yet there has been progress, because, with the failure of old theories, new more powerful ones have emerged. The search for true explanations in terms of the interpretation of thought will be similar to the scientific process. An interpretation will also be challenged in terms of new evidence and new argument. In the process of critical interpretation and reinterpretation of old and new evidence, a more accurate and truthful account of 'what really happened' will gradually emerge. However, even an interpretation which is well supported by relevant evidence will not be immune from critical attack and possible replacement.

Implications of Idealism

An important implication of the adoption of an idealist approach to explanation is the methodological separation of human and physical geography. This separation of the two major parts of the discipline, however, does not imply that human geographers have no need to consider the physical environment or indeed that physical geographers can ignore human activity. The separation does mean, however, that each component of geography will consider the content of the other from its own distinctive point of view. The human geographer will consider the physical environment mainly in terms of the way people of different cultures and circumstances have used its resources. As human ideas on technology, institutions and social priorities have changed so have the relationships between man and his environment. The physical geographer, on the other hand, is basically concerned with man as an agent of landscape change. Questions dealing with such topics as soil erosion, climatic change and water pollution are all amenable to scientific analysis and the special methods one needs for understanding human behaviour need not be invoked.

The idealist approach is premised on understanding the rational strategy adopted by individuals and groups to achieve their goals. Once a geographer has been able to demonstrate a rational connection between the geographical behaviour of a group and its thought (supported with appropriate evidence) he has succeeded in providing an explanation of their behaviour. This explanation does not rest directly on laws or theories. A single unique case is capable of being understood using the idealist approach, as well as an action that might be repeated a thousand times. The emphasis on the unique permits the idealist geographer to take full account of the great diversity of peoples and cultures in their many geographical contexts. Although the main emphasis of an idealist approach is on unique situations, this does not preclude, as I pointed out above, the formulation of generalisations. However, generalisation will follow rather than precede empirical work and will be taken no further than the evidence permits.

The importance the idealist method places on understanding human activity in its cultural context means that it is ideally suited to the study of regional geography (Guelke, 1977). In seeking to establish an initial classification of regions the idealist geographer will attempt to group together people who share a common culture or world view, because such views will largely shape geographical behaviour. Within the broad categories established subdivisions would be needed to take account of, for example, physical conditions. These subdivisions, however, would not be identical for each cultural region, because an attempt would be made to classify the physical world in terms of its resource potential. This potential would vary from place to place in relation to technical, social, political and economic factors.

Many human geographers have made use of what amounts to an idealist approach in man/land, landscape and regional studies. A study is considered to be idealist if one's understanding of the human activity described depends on one's seeing a rational connection between what people did and what they thought. Such a rational connection might often be established implicitly. A researcher who, for example, described the use of an area in terms of the human needs and purposes of its inhabitants might not explicitly regard his explanation as depending on rational understanding, yet to the extent that his account was explanatory it would, in the absence of laws and theories, likely depend on the ability of a reader to grasp a rational connection between what people thought and what they did. It is unfortunate that the idealist approach to regional study requires a greater appreciation of cultural milieus than can usually be expounded within the bounds of a general textbook covering a continent or major portion of it. In consequence, it is difficult for an author of such a textbook to establish a proper context within which the rationality behind the activities of human groups, within their distinctive milieus and traditions, can be properly elaborated. Nevertheless the idealist approach has often, if implicitly, been used by geographers to elucidate the human use of the earth.

Even geographers who would consider their work to be rigorously scientific often make implicit appeals to the rationality of human behaviour. Indeed, many of the *a priori* models and theories used by human geographers do not depend on empirical laws, but on understanding hypothetical situations in terms of rational principles. For example, the von Thunen model of agriculture is developed on a model of rational man. The user of this model comprehends the behaviour of hypothetical farmers at various distances from the isolated city in terms of what makes rational economic sense. This understanding is close to the idealist method of rethinking or *Verstehen*, and the idealist geographer has no objections to the use of simplified models as aids to understanding how human activity on the earth's surface might be affected by certain factors under ideal conditions. If *a priori* models are used in the explanation of real situations, however, their status becomes empirical and they must meet the normal requirements of any scientific theory.

The ultimate objective of geographical study is seen as the acquisition of understanding and insight into man's activities on the face of the earth. The idealist methodology is well suited to this purpose, because it enables human geographers to comprehend the meaning of such activities in their social and cultural contexts. In uncovering the meaning of an event or action the scholar is basically concerned with identifying the ideas which enable him to understand why an individual or group acted in one way rather than another. A satisfactory explanation is one that removes the puzzlement that a certain action or actions might have created (in the minds of observers and scholars) by showing that such actions were the logical outgrowth of a process of rational thought. The idealist investigator is forced to probe deeply into the social, political and economic ideas of the peoples he studies in order to be able to reconstruct their thought correctly. In this process there is a prospect that real insight into the activities of a people will be forthcoming. The depth and extent of the insight achieved will depend on the qualities of the investigator and on the acuteness of his questions. However, the results of his study, whether trivial or profound, will need to be supported by relevant evidence and logical arguments (Guelke, 1974, pp.200-2).

Idealism and Other Philosophies

The fundamental strength of the idealist approach in comparison to that of approaches based on the model of natural sciences is that it permits an understanding of the meaning of human geographical activity which is not dependent on the formulation of general theories. The scientific or logical positivist notion that disciplines lacking general theory are in a 'prescientific' or natural history stage of development forces those who share these basic assumptions to adopt a theoretical (scientific) approach or to be content with description. In geography the adoption of theoretical approaches has not provided a solid foundation for the causal understanding of human activity, because the empirically well-verified general theories and laws needed for successful scientific explanation of human geographical activity have not been produced. A somewhat paradoxical situation has emerged in which potentially valuable theories lack empirical status and well-established empirical facts lack theoretical explanations.

The attempt to apply the methodology of the natural sciences had particularly disastrous results in the study of regional geography. Collingwood, in contrasting the methods of natural science and history wrote:

the historian need not and cannot (without ceasing to be an historian) emulate the scientist in searching for the causes or laws of events. For science, the event is discovered by perceiving it, and the further search for its cause is conducted by assigning it to its class and determining the relation between that class and others. For history, the object to be discovered is not the mere event, but the thought expressed in it (Collingwood, 1956, p.214).

Regional geography conducted along the lines of the natural sciences encouraged its practitioners to investigate external relationships between such things as soils and settlement, climate and crops. In the absence of general theories and laws such relationships were basically descriptive not explanatory. As a consequence much regional geography was little more than systematic inventory and description, and many became disillusioned about its potential academic value. These problems are avoided with the idealist method which permits the researcher to progress beyond a description of events and relationships to the meanings expressed in them.

The behavioural and perception geographers share with idealists a concern with the subjective foundations of human behaviour. The procedures and aims of behavioural and perception geography, however, are quite distinct from those of the idealist. Behavioural and perception geographers are concerned with using rigorous, scientific procedures (many adapted from psychology) with the view to contributing to the development of geographic theory. Although the subjectivity of images is recognised, the aim is to develop procedures by which individual images can be extracted from their subjects and incorporated into theory in the usual way (Golledge, Brown and Williamson, 1972). The aims of behavioural and perception geographers are, therefore, entirely in keeping with the scientific emphasis of the modern spatial discipline. The idealist geographer's concern is with the meaning of thought in its context and is dependent on *Verstehen* or rethinking. The construction of formal, scientific theory is not considered necessary for the achievement of this aim.

The idealist philosophy has elements in common with phenomenology, which also emphasises the need for a non-positivist approach to the study of human activity. In particular, the phenomenologist, like the idealist, is concerned with the meaning of objects and events in terms of their human purposes. However, the phenomenologist does not make the sharp distinction the idealist does between intellectual and emotional life. Rather, the phenomenologist seems more concerned with providing a description of the life-world as it is actually experienced (Buttimer, 1976, pp.277-92). This concern avoids the abstractions involved in both positivism and idealism. However, in seeking to describe life-worlds there is a strong element of subjectivity. This subjectivity is inevitable, because the approach fails to distinguish those elements of human existence that are open to objective understanding and those that are not. Thus idealism differentiated from phenomenology both in its concern for objective, verifiable knowledge, and in its emphasis on the causal understanding of human activity as opposed to the descriptions of human life-worlds.

On a philosophical level idealism and Marxism have little in common. The idealist gives priority to ideas as the foundation of human existence and knowledge, the Marxist advocates a form of materialism. Yet these important and real philosophical differences are less marked in some empirical work. A Marxist geographer in analysing human behaviour often seeks to identify the interests that different groups have in specific situations. The result is an analysis along idealist lines in which it is possible to see a rational connection between behaviour and ideas. The idealist, however, does not subscribe to the theoretical ideas of Marxism, and where such ideas become dominant Marxist scholarship is considered to be of less value. The rejection of Marxism as a theoretical system is based less on its radical character than it is on its failure when subjected to genuine empirical tests. Any theory that is not, in principle, refutable must be considered a philosophy or ideology rather than a valid scientific theory with empirical status and explanatory power.

Although Marxism is rejected by idealists, Marxist geographers have made an important contribution to the discipline by uncovering some of the hidden biases of many positivist theories. A strict application of logical positivist criteria would effectively eliminate subjective judgements in geographic theory, but such criteria have never been rigorously applied in human geography. In consequence, much of the theoretical and statistical work of positivist geographers was not truly objective, but biased in favour of the middle class. Although Marxists were able to expose shortcomings of the conventional wisdom in geography, they did not solve the problem of bias. Marxist geographers were explicitly concerned to promote the interests of poor and working people in opposition to the middle-class bias of much positivist geography. The idealist is concerned to create a genuinely objective human geography by avoiding any theoretical and ideological commitments that might create a built-in bias towards one group or another. In analysing a specific situation the idealist seeks to understand human behaviour in terms of the ideas and beliefs of the people involved in it, and to refrain from making value judgements. Objectivity, however, is not something that can be easily achieved, and the level of success will ultimately depend on the integrity and imagination of individual scholars working in an open and critical environment.

The successful achievement of objectivity is made more difficult where human geographers are involved in applied or problem-solving research. It is unlikely that any set of recommendations can be made which would not benefit one group more than another. Although geographers will be in a good position to make recommendations about a situation if they have an appreciation of the objectives and ideas of those involved in it, they should be careful to make any underlying assumptions of their recommendations explicit to avoid becoming involved in the political process. In a world of conflicting interests there would seem to be ample room for geographers who (by being less involved in immediate and day-to-day problems) are able to gain a perspective on things that more actively involved decision-makers are unlikely to develop. This role would also seem to be consistent with the geographer's traditional concern with understanding specific problems in their regional and cultural contexts.

Conclusion

The idealist methodology makes a fundamental distinction between the explanation of human and natural events. An event involving human decisions must, according to the idealist, be explained in terms of the thought contained in it. The method of *Verstehen*, or rethinking makes it possible for an investigator to understand rational thought. In essence, rethinking involves discovering the human purpose behind an action; it does not involve the impossible task of re-experiencing the emotional life of others. The idealist maintains that it is possible to re-enact the rational thought of another person or group, because rational thought can be detached from the original action of which it was part and comprehended by the scholar.

The idealist approach to the explanation of human activity is premised on understanding a situation in terms of the points of view of those involved in it. A thorough knowledge of the cultural context and specific circumstances of individuals is needed to interpret their points of view correctly. This emphasis on the unique, however, does not mean the human geographer is tied to description, although careful description will be needed to place events in context. However, the crux of the idealist approach involves discovering why people behaved in one way rather than another by reconstructing their thought. The method of rethinking permits the human geographer to grasp the meaning of geographic activity in terms of human reason without having need of laws or theories, except, indirectly, in supporting roles.

The idealist approach is quite distinct from the other major approaches currently accepted by geographers. In contrast to the logical positivist, the idealist emphasises the autonomy of the social sciences, the importance of understanding human activity in terms of thought not general theories and the central role of uniqueness in human geography. The idealist shares with the positivist a commitment to objective explanations supported by empirical evidence, but maintains that such explanations need not be based on a model derived from natural science. Idealism differentiates itself from phenomenology by making a sharp distinction between an individual or group's rational thoughts and their emotions and feelings. This means that the idealist emphasises the objective reconstruction of thought as a basis for the explanation of human activity rather than the subjective description of 'life-worlds'. The idealist in contrast to the Marxist emphasises the importance of ideas as the foundation rather than the reflection of human material conditions.

The methodology of idealism is seen as providing an appropriate analytical tool for gaining fundamental understanding of human behaviour similar to the fundamental understanding that theoretical knowledge gives us of the physical world. In human geography this method is applied to the investigation of human activity on the surface of the earth. In understanding the thought behind human activity on the earth the human geographer gains a perspective on the ideas and values of the earth's various cultures and peoples. Such a perspective is considered to be of value in the general scholarly quest to understand the nature of human society. Although an idealist approach is essentially academic, in the sense that an idealist is concerned with understanding, not manipulation, the results of an idealist analysis are, just as in fundamental physical research, of potential practical value, because any attempt to change the world is more likely to succeed if it is based on an understanding of situations as they really are.

8 REALISM

Edward M.W. Gibson

Introduction

It is probable that the last decade will be regarded as a turning point in the writing of modern geography. Compared with earlier times, the number of important works advocating different philosophical departures was relatively great. Borrowings from phenomenology, existentialism, Marxism and idealism have been presented as new humanistic approaches. Each, it has been argued, could rescue human geography from the stagnating influence of naturalism, empiricism, positivism and instrumentalism. Of the main philosophical movements of our time, realism is one of the few whose virtues have not been advertised as a cure for geographical discontent. I believe we have need for an exposure to realist philosophy — although I would not go so far as to propose realism as an excelsior. If realist philosophy is not marketed for its overall virtue, it is valuable as a means of showing the limits of other philosophical wonder men now being promoted — particularly idealism, the tradition with which realism is commonly contrasted in our time. Certainly a balanced argument calls for a deeper consideration of the connections between geography and realism.

Realist Philosophy

In the early history of philosophy, particularly Platonic-Socratic thought, the term realism was used in opposition to nominalism for the doctrine that universals and abstract entities have real objective existence. But the same term can and does take separate forms. In our time realism is, in opposition to idealism (See Chapter 7), the doctrine that there is a world of physical things that exist independent of our perception and cognition of them. We also use the term beyond the narrow theories of knowing, perceiving and cognition. In the broader sense of a philosophy of science, it is a form of explanation. In opposition to other philosophies of science such as naturalism, positivism, instrumentalism and idealism, realism is based on the doctrine that human science is an empirically based rational enterprise which explains observable regularities by describing the hidden, but 'real', structures that causally generate them.

For earlier realism we use the term Platonic-Socratic or Scholastic Realism. Since the Enlightenment realism is qualified by a prefix, particularly when it is used in opposition to idealism. Direct Realism and New or Critical Realism both refer to theories of knowing, perception and cognition. Scientific Realism is used when we are discussing philosophies of science, forms of explanation and questions of metaphysics. These different technical uses of the term are important because they mean geographers are, or could be, either direct or scientific realists.

In the following paragraphs we will be analysing these different conceptions of realism and their connections with writings in geography. The discussion to be conducted will stress the contrast between conceptions of realism and the philosophical traditions with which they are in opposition at a point in time. But throughout these contrasts, we must keep in mind two observations, two problems that will face anyone who reads the philosophical sources used in this chapter. First, that at any point in time it is possible to see not only contrasts between two competing philosophical traditions but similarities as well. That is, for instance, we may see in the quantitative environmental studies of the last two decades characteristics and assumptions that properly parallel both Critical Realism and idealist traditions. Similarly, substantive historical geography may use explanations that combine assumptions that properly belong to positivist, idealist or realist philosophies of science without any bibliographical footnotes documenting conscious borrowings. We will note these parallels as tendencies in geographical writings.

There is a second problem geographers will face in reading the original philosophical sources. This concerns the survival in contemporary philosophy of earlier conceptions of realism, conceptions that are technically different from, but logically connected with, forms of modern realism. Thus although we must technically distinguish between Platonic-Socratic Realism and Scientific Realism, we must be equally aware that when we use realist explanations they are logically connected with Platonic-Socratic Realism. Both conceptions suspend belief in the 'appearance' of things and assign ontological, objective status to hidden, abstract entities. That is the nub of their logical connection.

Today the elements of a realist geography fit together into a humanist, yet theory-based, alternative to both positivist and idealist philosophies of geography. Each element in realist philosophy has been present in earlier stages in the history of philosophy. Therefore, so as to make clear possibilities and limitations of a more realist geography, it is useful to begin with earlier stages.

Platonic-Socratic Realism

In devoting space to a discussion of Plato we are doing what must be done to understand contemporary realism. That Plato and his medieval students are mentioned is not a matter of honorifics. The rational theory that the forms we see, touch, taste and smell in time and space do not exist and are not knowable with certainty by our senses was first put by Platonic-Socratic writers. It was in Plato's theory of ideas (Russell, 1959). It is perhaps best understood here by using the distinction between the particular mountains I see from my desk in Vancouver, British Columbia, and the general, the universal geographical term 'mountain'. The geographical term refers not to Hollyburn Mountain I can see or any particular mountain, but to the universal mountain that is separate from any mountain we see. The meaning of the universal mountain is not fixed in any time or space; it is eternal. The eternalness of the universal term 'mountain' is held to be based on an abstract 'ideal' mountain, an unchanging entity existing in another world, a non-physical world. I believe that were Plato here in Vancouver today, he would say that the Hollyburn Mountain we can walk to and touch is what it appears to be only because it is related to the ideal mountain. This particular Hollyburn Mountain will be worn down and carried to the ocean floor over geological time. The ideal mountain is rigid and fixed. This is what is meant when we say that particular mountains are only 'appearances'.

On the other hand nominalists, notably Aristotelians, denied the existence of an ideal mountain. For the nominalist of early philosophical history, 'mountain' was a mere term. What, and only what, is real is the particular mountain that we can all see and touch, including, of course, Hollyburn Mountain before me.

The battle between realists and nominalists over the existence of abstract entities, or in other words, problematic entities, carried over into the Medieval Period.

By the close of the Medieval Period, roughly what we have described as Platonic-Socratic thought became known as Scholastic Realism. No person is more important in developing Scholastic Realism than is John Scot. According to Clarence Glacken, medieval geographical writing was greatly dependent on scholastic realism (Glacken, 1967). Certainly, John Scot's essay 'On the Division of Nature' is based on scholastic realism. In this essay it is reasoned that the divisions of the physical world all signify something hidden. In themselves they are not real, so Scot believed.

The cyclic processes of the physical world – seasonal and astrological cycles for instance – all proved for Scot the existence of a divine order, a harmony and a law. They proved that the ordinary sense-world was not real. This appearance of the world was like 'fallen man', a reflection of a perfect abstract entity. Physical geography, in other words, was a consequence of a problematic order. Scot's debt to Plato was obvious. In 1877 John Scott lost his life because of his realist doctrine. But his doctrine was not taken with him to his grave. A traceable line of Platonic-Socratic Realism runs from Scot through Western thought up to our time. The problem for us is that some elements of this early form of realism become known as idealism while other elements continued in essentially the same form as qualified variants of realism.

Platonic Realism and Enlightenment Idealism

The confusion begins with philosophical developments of the Enlightenment. The word 'idealist' first came to be used as a term in philosophy in the work of Gottfried Leibniz (1646-1716). He used it to refer to those philosophers who, like Plato, held anti-materialist views in the broad metaphysical sense. But, by the end of the eighteenth century, when the narrow theories of knowing and perception were developed by the Enlightenment greats – John Locke (1632-1704), George Berkeley (1685-1753) and David Hume (1711-76) – the same term, idealist, was used. The overlap between the broad term Idealism or Platonic Realism and the narrow term idealism or the belief in a physical world dependent on mindfulness of it, is the core of our semantic problem. During the Enlightenment, philosophers who held idealist-like metaphysics and epistemologies became dominant in Great Britain. By the beginning of the nineteenth century self-claimed idealists become dominant throughout Western philosophy.

Direct and Naive Realism

Not too unexpectedly, we can look back into the nineteenth century
and find a resistance to the idealist movement, a polemic against
idealism. This movement, which owed much to the Victorian
philosopher Cook Wilson at Oxford University between 1899 and
1910, concentrated on theories of knowing the physical world
(Passmore, 1966). Cook Wilson labelled his theory, an argument
against Berkeley's idealism, Direct Realism. Students of Cook Wilson
followed him in denying the existence of any problematic or
abstract entity, a denial which of course runs counter to Plato. For
Direct Realists nothing existed that was not observable in time and
space. From this ontology they developed a logic of perceiving the
world, a common-sense logic which argued that our views of the
world are constructed in the mind by an interaction, through society,
with the physical world.

Direct Realism by itself appears to have had little impact on
geographical writing. But associated with this academic movement was
a popular movement called Naive Realism. This latter movement has
had a sustaining influence on geographical writing since the
Victorian Period. It is an influence that is most apparent in what
has been called commercial and military geography; but it is also
apparent in some forms of resource geography. Before we demonstrate
and assess this impact on geographical writing we need to appreciate
the underlying logic of Naive Realism.

For the Naive Realist, or common-sense geographer, the mind grasps
the world in a simple effortless process, something we do all the time.
True, our perception can occasionally go wrong: we fail to see the
'blue' hue to the distant mountains. Or the distance we walk between
the classroom and library 'seems' much shorter than the two hundred
yards confirmed on the campus map. Still, these are minor exceptions
that we discuss because of our inattention. No doubt, with the
collaboration of others, with the aid of improved instruments, we can
establish the 'real' nature of the physical world. Geographical facts
of observed phenomena and changes within them can be objectively
established and any question of unseen entities, problematics,
abstract forms or subjective impressions are irrelevant. With due
precaution, so say common-sense geographers, we can know a place
as it really is, as the topography looks, as the soil feels, as the water
tastes and so on. We may understand that the measures we use in
objective studies, like two hundred yards for example, are products

of our minds and not the external world alone. Yet we also understand that they are not the products of private illusions either. Each measure is neutral and exists independent of our thinking so far as Naive Realists are concerned. Thus, according to this logic, conveying the world requires nothing more than a description of its facts and most assuredly it does not require assumptions of 'hidden' entities.

Most of us have read geographical works which, although not directly stated, seem to share the logic of Naive Realism. Those who find it necessary to condemn refer to such writing as 'capes and bays' geography. The intentions of common-sense geography are commonly those of social reform or national survival and blanket condemnation of their work seems less than warranted. Consider for instance the work of Dudley Stamp (1889-1966) (See James, 1972). Seeing the need for natural resource inventories in economically depressed Britain, Dudley Stamp used secondary school students to produce a factual inventory of British natural resources mapped at one inch to the mile between 1931 and 1935. In the years of adjustment following the Second World War, the same tendencies toward Naive Realism provided a basis for the national land-use mapping programme of the Government of the Dominion of Canada, a programme that created an increased demand for qualified geographers and which ultimately helped to create new departments of geography like those of McMaster University in Ontario, and the University of Manitoba. Even today, with the wave of geographical studies in urban and regional planning, there remains a tradition of Naive Realism. Here we have in mind resource inventories.

Naive Realism has perhaps never been so prominent as during the two world wars of the twentieth century. Geographers used by the United States of America during the Second World War definitely appealed to Naive Realism, not to 'problematic entities'. According to Preston James, the United States Government used geographers to good effect to survey commodity imports and exports and plan war strategies (James, 1972).

The tradition of Naive Realism presents a paradox in our time. The tradition, after all, did create a class of professional geographers: it was the basis for exuberant expansion and higher status for geographical studies in more and more universities. But there were problems. Not the least of these was a demoralisation of the kind that comes from mechanical inquiry when it is cut from the ongoing philosophical debates surrounding it. The consequences of the extremes of Naive Realism was the so called Quantitative Revolution of the 1960s.

That this revolution was based on a move toward philosophical Positivism has been clearly argued (Harvey, 1969). What is not so clear is that it was also based on the twentieth-century tradition of Critical Realism.

New and Critical Realism

New Realism developed around the writing of T. P. Nunn (Passmore, 1966). The heart of Nunn's theory was his refusal to admit that anything we experience depends for its existence upon the fact that it is experienced. In other words, the objects of our perception are actual properties of the physical world. These objects of our perception, as opposed to the objects of the world itself, are called 'sensa' or 'sensa data'. With the appearance of a book titled *New Realism* in 1912, a book with contributions from both sides of the Atlantic Ocean, the seeds of a revitalised Realism came to flower.

The central teaching of the essays in *New Realism* was neutral monism, that is, nothing is admitted to exist excepting objective 'sensa'. To know the world, if one is a New Realist, is to know the relations between these 'sensa'. There is a marked empirical spirit to this teaching, a spirit with which few philosophers of our time would not wish to be identified. Out of such spirit Samuel Alexander wrote *Space Time and Deity* in 1920 (Alexander, 1920) and Alfred Whitehead completed *Process and Reality* in 1929 (Whitehead, 1929). Alexander's work shows the affiliation between new realism and neutralism, a main tradition within which nineteenth-century intellectual thought (including human geography) developed. 'The temper of realism', Alexander says,

> is to de-anthropomorphise; to order man and mind to their proper place among a world of finite objects; on the one hand to divest physical things of the colouring they receive from the vanity or arrogance of mind, on the other to assign them along with their minds their due measure of self existence (Passmore, 1966, p.269).

Although it is difficult to demonstrate it by pointing to specific footnotes or text, the large debt that early twentieth-century geographical determinism owes to critical realism is plain enough; read Alexander's *Space Time and Deity*, then read Ellsworth

Huntington's *The Mainsprings of Civilization* written in 1931
(Huntington, 1945). To the critics, the unscientific methods of
Huntington were an opportunity to display their powers of exegesis
and to top other critics who might have otherwise given up critical
realism and naturalism as a hopeless philosophy for human geography.
Hope was taken in a form of critical realism as a theory of perception.
In 1938 the British philosopher Dawes Hicks published his manifesto
against idealist doctrines of knowing and perceiving, doctrines such
as those held by his contemporary, Robert Collingwood (Hicks, 1938).
Hicks' perception theory is a refinement of the sense-data theory of
Bertrand Russell (Russell, 1959). It denied that the perceiver is ever
directly aware of physical objects existing independently of him or
her; and at the same time it held that he or she can derive knowledge
of independent physical things from sense-data. Sense-data comprise
ordinary language such as 'blue' hills, or scientific observational data
such as $20°C$ temperature. The physical world or the mental world
is not knowable but mentally produced data of the separate physical
world are. In the view of critical realists it is these data with which,
and only with which, science can deal.

Although they do not claim as much, since they referred back only
to the psychology of their time, the geographers who pioneered in
environmental perception and behavioural studies ultimately based
their logic on critical realism. Some subordinated this realist logic
to positivism and hence observational language; other pioneers
retained a bias toward ordinary language and an allegiance to the more
traditional inductive and empirical methods of geography. As William
Kirk, the pioneer of this latter viewpoint, put it in 1950, 'this
internal [that is psychological] environment we may call the
"behavioural environment" and in this environment the gap is closed
between Mind and Nature' (Kirk, 1951, 152-60). Had Kirk said the
behavioural environment represents the eternal world of Nature, he
would have connected geographical studies in perception to
idealism and phenomenalism. But he did not. The difference is
important if we are to be accurate in the relationships between
geography and philosophy.

The case of John Wright is admittedly less straightforward, in part
because Wright published more than Kirk. He published from different
philsophical viewpoints. Behind the barrier of inductivist-empirical
methods John Wright romped freely through historical geography
much like Robinson Crusoe exploited his Lost Island. In his essay
'Map Makers are Human', the logic is grounded in idealism and dualism.

The object before the camera draws its own image through the operation of optical and chemical processes. The image on a map is drawn by human hands, controlled by operations in a human mind. Every map is thus a reflection partly of objective realities and partly of subjective elements (Wright, 1966).

In other places the logic of his writing seems closely tied to a critical realist approach to ordinary language. Geography to him was like history: it 'should be studied empirically and inductively, the facts [sense-data] being allowed to speak for themselves' (Wright, 1966). Critical realism indeed!

Untethered intellectual Robinson Crusoes are numerous among historical geographers outside Europe. But recently some geographers are securing guidelines between historical studies and idealism. The primary motive for this is some orderly philosophical base for the discipline. If I understood what Leonard Guelke, the most published advocate of idealism in geography, is arguing, we have developed methods of entering the minds of our subjects so as to think their thoughts and justify their expectations, methods that will determine human intentions and understanding of our role in changing the earth (Guelke, 1974, and Chapter 7). In fact the methods idealists employ do not bear upon all the problems confronting geographers: they are limited to problems arising out of language. For the idealist it is essential that we know what it is like to be human, then we may write human geography. Since it is impossible to know what it is to be a geological formation or a river valley there can be no historical geography of their development. An entirely idealist geographer would have to limit his or her data solely to historical documents, artefacts or field notes from participating observers as an account of what happened. An idealist would omit other types of evidence. He or she would thus leave out evidence from non-human Nature and evidence from the present 'looking back'. But realist methods employing ordinary language would pass all these types of evidence, including written evidence, through its turnstile. The realist geographer finds it entirely proper to interpret human uses of the earth through a dialogue between the investigated and the investigator. For the realist, as opposed to the idealist, the deliberate suppression of modern theory and evidence to 'become' the person being investigated, is less than satisfying. You need to go beyond idealism. Agreed that you require a sympathetic attempt to enter the thought of a person investigated, but you should never abandon evidence developed from subsequent study (Keat and Urry, 1975).

To be fair, idealists do mention that they could modify their assumptions in the face of something like medical evidence. Yet this construct presupposes a fundamental dualism and not merely a pseudo-problem of geographical methods. To modify idealists' claims that human thought has a life of its own, independent from material and natural processes, to accommodate medical or geological evidence would indeed trivialise the traditional distinction between critical realism and idealism. Once you admit to evidence external to the mind for the evaluation of past uses of the earth, you become a critical realist.

If there be confusion between idealist and critical realist theories of perception, there is certain to be at least that when it is said that critical realists can be positivists (logical positivists). This depends on the form of explanation geographical studies take. It mainly has to do with perception and behavioural geography. Where such studies – the work of Reginald Golledge for example – limit evidence of sensa to only quantitative indices of the mind and physical world, they clearly belong to the tradition of critical realism (Golledge and Amadeo, 1968). Where the same studies then arrange these data into hypothetical relationships derived from theoretical laws and proceed with tests for verification (Cohen and Nagel, 1934) or falsification (Popper, 1961), they also plainly appeal to the philosophy of science we know as positivism. The outstanding characteristic of geographical literature of the 1960s, the 'new' geography, was this potent concoction of critical realism and positivism.

Scientific Realism

In addition to being a logic of behavioural and perception studies, realism or, better still, a particular tradition of realism – scientific realism – can be used by geographers as a framework for explaining things. In fact by the evidence of recent 'key words' in geographical circles, the spirit of scientific realism will be a hallmark in the coming decade. We know how, on the basis of geographical hypothesis, observed physical and social patterns can be traced to previously unsuspected origins. For instance, the spatial pattern of mortality rates for respiratory diseases can be traced to spatial variations in exposure to high radiation levels; or spatial patterns of soil types can be traced to geomorphological formations of early geological and climatic conditions. The habit of seeing mysterious (but predictable) connections of these kinds has been instilled even in high school geography students. Unsuspected 'causes' or 'antecedent conditions' are almost casually identified by students and professors of geography who analyse archival and survey data with a wide range of techniques and instrumentation (Harvey, 1969). On the other hand, the most philosophically trained geographer could rarely be convinced that our explanations in geography are altogether free of fantasy. This does not seem to matter; whatever be its status among the intellectual traditions of geography – positivistic, idealistic, phenomenological or realistic – explanation is an aspect of geography today. In the sense of which we are speaking, realistic explanations are largely to be understood in the context of competing intellectual frameworks. More than anything else, realistic explanations compete with positivistic and idealistic explanations.

Whatever we say about positivism and realism, in practice we concern ourselves with the characteristic questions each asks and answers. The heart of the difference in questions is that positivists ask 'how' a pattern is produced while realists ask 'why'. For the realist to answer the question why it is necessary to go beyond the logical argument of the positivist, to go to a description of the causal processes generating the observed patterns of physical and social regularities. But there is more to it than that for, while the positivist speaks of the logical (most frequently mathematical) necessity of the regularity, the realist speaks of *natural necessity*. Realists seek to discover the relational cause-and-effect in nature and society. As one realist philosopher, Rom Harré, puts it, at least one of the conditions, causes or factors in an explanation must refer to a thing that can be pointed out in the natural world, a thing that is defined in time and space (Harré, 1970). This latter point is significant for distinguishing between realism and idealism. As idealists will be quick to say, they too ask 'why' and not the question 'how'. While realists answer 'why' with reference to natural necessity and things in nature, idealists answer with mentalistic data about nature (or society). That is they answer not with reference to the doctrine of natural cause but to the doctrine of *Verstehen* and psychological processes.

The saying of Robert Collingwood, if it was Collingwood, 'The fact that certain people live, for example, on an island has in itself no effect on their history; what has an effect is the perception people have of that position', is rejected by realists. For the realists the importance of mentalistic data is limited. In fact realists explain regularities by reference to existing entities but not necessarily to entities perceived and cognised in time and space.

The central concern of realists with the reality behind appearance, with problematic structures and mechanisms, demands the use of theories and models in realist explanations. Models are virtually the same as theories for realists; both are used to attempt to describe causal structures which are incapable of being observed with any mechanical instrument such as a telescope or a chemical test. The procedure for realism is to draw upon familiar sources and construct models and theories that correctly represent causal structures in a manner open to empirical testing, then test the model as a hypothetical description of actually existing entities. So, similar to the positivist and opposite to the idealist, the realist explains deductively.

How then does the use of theory differ between positivism and realism? At the root of the difference is the source of the theory that for the positivist could be purely a logical conjecture, an abstraction; a mathematical law for instance, while for the realist there is most commonly an emphasis on a 'common-sense' argument following from the analogy (Hesse, 1966). In other words, positivist theories come from theoretical laws – laws without ontological status whereas realist theories come from problematic entities with assumed ontological status.

Realism exposes its opposition to idealism and its affinity with positivism, most openly in the matter of objectivity v. subjectivity. Both positivism and scientific realism believe in 'objects', a physical world which exists independent of our thoughts about them, a physical world that science can genuinely discover. Realist geographers would hold to this belief in the value of science in a kind of absolutism.

During the last decades we have stood in the shadows of relativism: Thomas Kuhn (Kuhn, 1970b) and Paul Feyerabend (Feyerabend 1970), are household gods in geographical discussions of philosophy. As bowing before household gods is a magical substitute for thinking about the problem of logic and meaning in geography, so is bowing to text-book visions of the future.

Postscript

We have seen that the historical connection between realism and geographical thought is not simple: today we give little attention to this philosophical tradition. Before the Enlightenment Period, Platonic-Socratic Realism showed a strong influence on the literature about the natural environment. Since the Victorian Period, direct realism, as a theory of perception and knowing the world opposed to idealism, has become the logic behind a persistent tradition of empirical description in geographical writing. The question of realism and geographical thought during the last three decades is one of critical realism and its consequences for both inductive and positivist work in geography – especially in environmental perception and behavioural branches. We have also outlined a use of scientific realism as a form of explanation that contrasts with idealist and positivist forms.

Our argument is not one of casting out other philosophies or even anti-philosophies from geography, but of balancing – most importantly between theory and non-theory and between anthropomorphism and naturalism in geography today. No such conclusions were in mind in 1978 when the idea of this chapter began to grow.

The editors of this book originally asked me to contribute a chapter on phenomenology and geography – instead I recommended someone far more knowledgeable about this topic, Edward Relph. I also asked if they would consider my developing one chapter on Realism. Philosophical discussions by geographers always contain 'problems', by definition. This discussion of Realism and geography, however, is double trouble. Little, if anything, was published on the subject and my reading had only then, early in 1978, begun. The need to know more about realism came from my work with two graduate students, Robert Galois (1979) and Alan Mabin, who were beginning Ph.D dissertations in historical geography and who wanted to employ the methods of historical materialism in their dissertations. Questions of epistemics and explanation that were theory-based non-idealist and non-positivist, in other words, questions of realism, have therefore guided my own reading and teaching over the last few years. But these same questions also come from an uneasiness I had about the basis of my own research in regional art and history and its connection to historical geography. There seemed to be no basis in cultural or historical geography that was parallel to Erwin Panofsky's form of iconological analysis (Panofsky, 1972) or the social function analysis of Alan Gowans (Gowans, 1974) and Norris Smith (Smith, 1967). Each of these was avowedly anti-idealist, each theory-based and logically belonging to the tradition of realism. Criticism of today's empirical and inductive historical geography founded upon these sources of art history issued edicts. But without deeper reading in philosophy the edicts were riddled with clichés. The challenge to reinforce one's work from the attacks of geographical oblivion was there. Having taken the challenge I soon discovered that realism was not quite the philosophy I had expected. It was plain in the first dark months of reading that help would be needed and I am indeed grateful that Milton Harvey and Brian Holly supported me. Had they not given me the advantage of reading Leonard Guelke's excellent chapter on Idealism before submitting a second draft of my work, many issues between idealism and realism would be unresolved here. Of that I am certain. I also benefited from reading Dereck Gregory's *Ideology, Science and Human Geography* (1978), which influenced my second draft. The upshot of this was a substantial rewriting of my section on Realist explanations, so that the knives of fewer critics will flash against that part of the chapter. If other parts of the work are understood, they in part owe their clarity to Michael Eliot-Hurst who read my first draft, tracked down loose lines of argument and suggested the organisation of the introduction as it now appears. But as the author it is I who must of course take full responsibility for the interpretations drawn.

Of what possible value are these interpretations? Any geographer worth his or her salt would have to agree on one value, regardless of whatever else may be thought. More attention to realism in geographical studies carries the promise of unity in geographical studies today. The 'new' geography revolution of the past three decades has vapourised. Only a minority believes that fundamental achievement and a unified discipline await a positivist geography; or that the discipline will be shaken by the power of phenomenology, existentialism or idealism. These latter philosophies cannot contain physical geography.

As the smoke blows away and the philosophical fragments settle in the next decade there is a strong force to break up geographical study into 'physical' and 'human' geography (Worsley, 1979). Nothing could be more destructive of the great achievements of geographical thought, of the intellectual traditions that are needed in these years of assault on the natural environment by post-industrial institutions, than whining about separating geography into two disciplines. It should be clear that a charm of realism is its capacity to shelter both physical and human geography under a single epistemological roof

(Eliot-Hurst, 1980). So housed, geography need not be anti-humanistic and not anti-naturalistic. It would employ theory, empirical testing and at the same time provide an analysis of the interpretive process of the physical and social worlds through linguistic and conceptual meanings. True there will be an asymmetry in a more realist geography. The explanation of social reality will require a different kind of understanding than the explanation of physical reality. But within scientific realism as defined here this obstacle can be surmounted without sweeping out either humanism or naturalism. To be engaged with the philosophy of realism can liberate geography as scholarship by eliminating the need to ask the questions do the physical and social worlds exist and can we know them objectively?

They do. We can.

9 ENVIRONMENTAL CAUSATION

John E. Chappell, Jr

Introduction

In the long history of ideas about the relations between human beings
and their natural environment, it has been assumed by most thinkers
from the very beginning, from the time Hippocrates defined health
largely in terms of harmony with external nature, that this external
natural environment plays a significant role in determining the forms of
our culture and their variations over space and time (Glacken, 1967,
pp.80-115). The leader in the organisation of academic geography in
the United States in the early twentieth century, William Morris
Davis, also believed very clearly in the importance of environmental
influences (Davis, 1906).

 · The fact 'that environmental causation has become an unpopular
concept since the time of Davis reflects more of our subjective
perceptions in an artificial world than it does of ultimate reality. As
Kuhn (1970b) has emphasised, scientific paradigms are often accepted
for psychological, aesthetic or other non-empirical or non-rational
reasons. In our own era, we have only recently experienced a
quantum leap in technological mastery over the natural environment:
first the Industrial Revolution, and then its electronic and cybernetic
elaborations. It is all too easy to persuade ourselves that natural
limitations have receded into the background and do not affect highly
civilised peoples in the same way they do the underdeveloped.

But who is more dependent on the environment in the long run:
the Kalahari Bushman or the Australian Aborigine on the one hand,
who scrapes a bare existence from the desert and must learn virtually
to smell the existence of water underneath the ground in order to
survive; or the citizen of a modern industrialised city, whose world
would come crashing down if imports of food and fuel were cut off,
and who probably could not survive alone in a wilderness for more
than a few days? Even Brunhes, whose main emphasis was on human
abilities and choices rather than on environmental limitations,
cautiously observed that it is a mere illusion to think that civilised
man has cast off the tyranny of the earth; he has merely made a
'more exact and . . . more Draconian' contract in order to become
more productive and specialised (Brunhes, 1920, p.614).

Accordingly, we have seen world population rise precipitously in the last two centuries, a phenomenon coincident in time with, and perhaps very closely dependent upon, the use of fossil fuel reserves. This simultaneous increase of people and of fuel use is called 'Hubbert's Pimple', or perhaps better 'Hubbert's Bubble', after the leading energy expert and predictor of the energy crisis of the 1970s, M. King Hubbert. Like a bubble, this curve on the graph can burst. In fact, the fossil fuel increase is sure to burst, and unless some extraordinarily swift and determined conversion to alternative fuel sources is accomplished within a few decades, the population aspect of the bubble may burst as well. As Brunhes observed, we have become dependent on nature in a very exacting and dangerous way.

This and almost every other manifestation of what is now recognised as a growing crisis of resource availability is most often interpreted as the result of mankind's carelessness. If only we would not pollute it with sewage and industrial wastes, Lake Erie would be clean and the most desirable fish would thrive in it. It is our fault, the experts groan. Admittedly it makes sense to concentrate on those problems which human action can ameliorate. But it is logically and empirically unsound to overlook the fact that nature is writing the rules in all of these games. Men may plough the soil and plant row crops on it and thus expose it to driving thunderstorms and massive erosion; but nature, not mankind, inexorably applies the laws of motion and of gravity and of fluid behaviour, which enter the door of opportunity for destruction that mankind has opened. Men may try to overcome a raging river with dams and levees, but nature decrees that each reservoir will eventually silt up, that much rain will fall downstream from the lowest dam and cause a flood anyway and that in other ways society would be better off if we tried to work with nature rather than against her in controlling the river (Kollmorgen, 1954). Similarly it is nature which has decreed that fossil fuels, at least for the next 50 million years or so, are strictly limited; men may learn how to squeeze more of the available supply out of the ground as they improve their technology, but they cannot alter the total supply. Even if we invent new technology which effectively changes the meaning of what is a resource, as we like to credit ourselves with being able to do, nature determines how much of the new resource is available.

The Moon has been reached, and we are now certain that it contains no grasslands to plough and plant. Probes of the nearby planets have also dispelled illusions that they might be habitable in the same sense the earth is. Dreams of discovering other planets in other parts of the galaxy, perhaps by some fantastic 'slowing down' of time and of our life processes while travelling in space, remain no more than bizarre flights of surrealistic fancy. We are stuck where we are, and sooner or later the grim reality of the declining ratio of resources to human population is bound to wake up even the ivory-tower theoreticians of man-land relationships to the fact that environmental limitations are quite important after all. As Rostlund (1956) so boldly and cleverly observed in the midst of contrary opinions all around him, the general educated public seems to have known this all the time.

Perhaps the great and unprecedented breakthroughs in science and technology such as Western man has achieved in the last 400 years could never have been reached without turning one's mind away from the thought of limitations on human ability; and now once these achievements are ours, it is only natural to consolidate our position on a new and higher plateau of ability and understanding on which our limitations must be redefined. On the other hand, perhaps only the less astute beneficiaries of earlier discoveries ever imagined that those who gave us these seminal ideas ever really took their minds off natural limitations, whereas the discoverers themselves actually heeded Francis Bacon's precept that 'nature can be conquered only by obeying her' (cited in Chappell, 1967, p.206). Our newly found power over nature, along with all the limitations and destructive side-effects it entails, does not, after all, rest on human laws we have created, as much as on natural laws we have discovered. Almost by definition, then, they could not have been obtained by ignoring natural processes and environmental limits.

Further discoveries in science and technology — and many remain to be made — will not be achieved that way either.

General Principles

In attempting to return to a more realistic evaluation of environmental limitations, several common pitfalls need to be avoided. To this purpose I now introduce a series of ten guidelines, each designed to counteract a particular fallacy common in anti-environmental-causationist literature.

(1) The 'single-factor determinist' is merely a straw man. Critics of environmental influences frequently set up such straw men in order to enjoy a hearty laugh and to clear the way for some declaration to the effect that only human choice and effort really matter very much

in the long run. But the real world of environmental theory is another matter. Only a few unimportant thinkers have ever been so careless as to claim that only one kind of cause makes the world as it is. The significant theoreticians were always more circumspect.

For example, Ellsworth Huntington, who made the most decisive step since the time of Hippocrates towards something new and conclusive in environmental-causationist thinking, managed to integrate very centrally into the scheme of his thought the so-called 'Gilfillan thesis', which is a form of cultural determinism. S. Colum Gilfillan, a sociologist some years Huntington's junior, argued that the centre of European civilisation moved from the Mediterranean to Northwest Europe around the sixteenth century because of technological breakthroughs in minimising the influence of cold, which theretofore had been the chief limiting factor (to use current ecological jargon) in preventing the full utilisation of the stimulating influence of the variable cyclonic climates. This idea is essentially equivalent both to the well-known concept of 'sequent occupance', and to the popular concept of relativity of resources to culture, already mentioned above. The central and most variable portion of the cyclonic storm belt is the resource. It could not be utilised fully until mankind learned how to fell its forests, build warmer houses, preserve food in moist climates and make cheap windowglass so as to read and work efficiently in cold winters (Huntington, in Van Valkenburg and Held, 1952, pp.208-19). Climatic change receded into the background in Huntington's mind as an explanation for this highly significant shift, compared to Gilfillan's insight.

Unfortunately, the unwarranted scorn of the geographic profession has obscured this argument from the light of respectability. Thus when an environmentally-oriented historian like McNeill tries to answer the same question about the northward movement of civilisation, he finds no intellectual shoulders to stand on; and he produces another and unnecessarily incomplete list of inventions, with an explanation less clear-cut and convincing than Huntington offered two generations earlier (McNeill, 1974).

Huntington's third causal factor, heredity, he applied much less astutely. This is not the place to get into a lengthy discussion on the pitfalls of genetic causation. Suffice it to say that even in the limited sense in which Huntington employed it, trying to explain differences between kiths or sub-racial groups rather than between races, it often lacked the firm empirical support he applied to other varieties of causation, and in fact often conflicted empirically with them.

Other environmental determinists also usually showed considerable appreciation of the causative power of human culture; both Glacken (1967) and Kriesel (1968) have demonstrated as much in the case of Montesquieu, for example. And on the other side of the fence, we can learn from T.W. Freeman's excellent survey of the career of Vidal de la Blache (Freeman, 1967, pp.44-71) that this founder of the possibilistic school of geography was very much aware of the frame of environmental limitations in which human choice took place; only his intellectual descendants seem to lose track of the frame. Instead of possibilism we now confront most often a kind of cultural determinism, which wants to ignore the environmental constraints as fully as it believes environmental determinism, in the typical straw-man version, ignored culture.

(2) The fact that all nature is one does not imply that unity cannot be analysed. Ecological awareness has taught us that all nature is interconnected; that you cannot affect one part of nature without affecting others (Commoner, 1971). We have been warned how difficult it is to discern whether a given feature is natural or cultural in origin. Many of the same commentators who impatiently reject the single-factor explanation then proceed to claim that it is intellectually naive to dichotomise or segregate the world into components. Unless causation itself is denied, this line of thought leads toward another kind of single-factor determinism, in which the possibility of several different types of causes is dispelled by lumping everything together.

Reasonable attention to intellectual history serves to expose the weakness of this position. It happens that our entire Judaeo-Christian intellectual and religious tradition which led to this great technological age and our unprecedented degree of power over nature is based on dichotomising reasoning. Sharp contrasts must be seen where nature offers them or her secrets will not be discovered. If that statement sounds too ethnocentric to some sensitive readers, let them suggest any other achievements in dominating nature, be they from Muslim, Hindu, Chinese, Mayan, or whatever cultures, explain how they have been brought about and see whether it ever could have been possible without dichotomising. The very number system which we employ to count the analysed parts came to us ultimately from India (or was it from China, as some guess?). Everywhere on earth, except in some few cases where unclear concepts are unduly glorified out of a fascination with novelty for its own sake, knowledge comes largely from dichotomising, from analysis. Look at your compass needle and then try to say that nature is all one, without any important polarities!

The confused and uncomprehending mind that cannot see the distinctions, perhaps because nature does not supply enough information to clarify them, should not be itself confused with external reality. It always is possible to distinguish differences in nature, even though our feeble minds may not accomplish the task. What we know at present is only a small part of what can ultimately be known. What is, is one thing; what we know, is inevitably another, a much smaller thing. It is not the reasonable implication of Heisenberg's Uncertainty Principle that particles do not have both velocity and position at the same time, but only that human beings cannot measure both at once.

The practical wisdom of analysing nature into its component parts is illustrated in the second of the reports on the world resource crisis sponsored by the Club of Rome. Many critics of the initial report, *Limits to Growth*, done by the Forrester team at MIT, argued that it was unrealistic to expect the entire world to suffer equally from resource limitations. Accordingly, a more realistic view was taken in the second report, by Mesarovic and Pestel (1974). They divided the world into ten regions, not all of which, it was estimated, would approach crisis simultaneously. In short, all parts of the world may indeed be interconnected, but this does not mean they are all alike. It is in fact the special purpose of geography to demonstrate how they differ, and why.

(3) Causation can indeed be discovered without specifying every link in a mechanistic causal chain. The weakness and ambivalence of the current cultural-determinist orthodoxy is nowhere better illustrated than in the treatment of the concept of causation. Some even have denied that the concept of cause itself is philosophically sound (e.g., Sauer and Leighly, as discussed in Chappell, 1969). The problem is not limited to geography, either. Two years ago I heard similar remarks from a member of the Forrester systems analysis team at MIT who seemed to believe that the concept of the feedback loop made causes practically indistinguishable from their effects. But computer programming and systems analysis are not changing reality so fundamentally as that. Recently I heard a more reasonable interpretation from Jay Forrester himself, who believes that causal chains do not disappear, but instead become or seem to be more complex, in the systems-analysis approach. If this complexity is

faced realistically, the approach becomes a useful tool.

Hume, the sceptic *par excellence* in the history of modern philosphy, sought to debunk the idea of inherent cause by pointing out that causation is always empirically determined, and that we never know more about it than that cause and effect are conjoined in space and in time, with the cause preceding the effect. As I argued a decade ago, however (Chappell, 1969), we can accept this definition of cause, and yet in so doing, show that 'mere correlations' are of the same qualitative order as causes. Thus we can use the argument of a sceptic to answer other sceptics. Of course we would like to bring the correlations closer together in time and space than is presently possible for, let us say, the influence of solar cycles on economic cycles, an example of a type of bold generalisation which was by no means restricted to the writings of Ellsworth Huntington. It began in 1801 with astronomer John Herschel (see Chappell, 1971), continued through the work of the economist W. S. Jevons in the late nineteenth century, and is being revived again in the suddenly thriving interdisciplinary effort to interpret history in terms of climatic cycles and influences (marked most notably by the International Conference on Climate and History at the University of East Anglia in Norwich, England, in July 1979). And the fact that links are still missing in this chain is no more reason to toss it out as invalid than we have reason to toss out Darwin's theory of evolution by natural selection because we can retrieve only a tiny fraction of the specimens needed to show every step of descent from ape-like creature to mankind.

And speaking of Darwin, I find it amusing that some of the same sceptics toward environmental causation who cry out for more definite and complete 'mechanisms' have been known to debunk Darwinism because it is 'too mechanistic' a theory to apply to human culture, with its causal relationships of living things to the natural environment. Then to further contradict themselves, they may turn around and heap praise upon the 'gravity model' as applied to cultural diffusion; in other words, analogies to biological processes are too mechanistic, but analogies to gravitational attraction between inanimate bodies are all right. In reality, of course, all that we know about natural processes seems mechanistic in comparison with the mysteries we have not yet solved. The mystery of life is likely never to be fully unravelled by science; but that is no reason to reject what little of it we do understand in terms of mechanisms.

By the way, do not imagine that natural selection alone explains evolution; it says nothing about the origins of the variations on which selection operates, nor is there any fully satisfactory answer to explain these variations yet, except in terms of 'chance' mutations brought on by impinging radiation or by random chemical combinations within the genes. But 'chance' in science is little more than a word for what we have not yet understood. It is only a myth that quantum physicists have shown the universe to be governed by chance; they have instead merely emphasised how much we cannot, or cannot easily, find out about the universe – and a number of physicists interpret the situations this way themselves. As for Darwin, his disciples have evaluated the concept of chance to become one of the cardinal tenets of their faith. But Darwin himself regarded chance as merely a cover word for still unknown natural laws. He denied ultimate purpose in nature but not causation and determinism (Greene, 1959, pp.302-7). Even so, there is nothing inherently purposeless about Darwinian evolution, especially if one interprets 'purpose' broadly to include, for example, the goal of coming into harmony with one's environment.

(4) The most profound truths are usually the hardest to discover, to confirm and to quantify. Critics of environmental causes may complain that they could not be real or we would know more about them by now. On the contrary, the little effort that recent science has put into them should not be expected to yield much in the way of results. The few theories that have been reasonably well confirmed, such as of the influence of variable weather on energy and creativity, are enormously productive if only they are put to use. Because precise data cannot be found to quantify every aspect of these influences is no reason to turn our back on them and be content with solving smaller problems. There are large, important bodies of thought, such as our religions, where quantification is very limited by the nature of the subject.

The computer seldom deals realistically with the largest problems, because the largest problems cannot be prepared and fed into artificial computers properly. But the most elaborate computer at our disposal is still the human mind despite its tendency towards subjectivism. We should trust more in the considered judgement of the well-trained and well-informed mind than we usually do. It is simply not true that nothing should be believed before it is fully quantified. In fact, as Kuhn (1961) demonstrates, many important scientific paradigms were accepted long before they were quantitatively confirmed, for example, Newton's law of gravity more than a century before. The quantitative confirmation occupies a considerable proportion of the effort of 'normal scientists' during the life of the paradigm. The original acceptance was probably based on a limited glimmer of this confirmation, but not on everything that was needed to be fully certain about it.

The trouble is not that quantification is somehow bad, or inherently misleading, but that very large problems do not come equipped with neat tables of data selected so as to yield definitive solutions. Which faith, or which ideology, should we believe in? Should we wage war for this policy goal, or that, or not at all? Numbers can be found to use in answering such questions, but rarely do they fully cover the ground. Who is to choose between conflicting sets of numbers, and determine which are significant?

Let us consider examples from the greatest crisis in twentieth-century American history. Did the North Vietnamese kill 500,000 enemies after taking power, as President Nixon assured the nation over national television, or only about 10,000, mostly in connection with a collectivisation campaign, as the research of Gareth Porter later revealed? The difference does not justify killing even the lower total, but it may be enough to determine whether or not the American people will be so shocked as to seek destructive revenge on behalf of the victims. Also, did hundreds of thousands flee to the south after Ho Chi Minh took over? And if so, why? Were they afraid of a cruel despotism? Or did they fear accusations of treachery because they had become Catholics, or in some other way been associated with the crude, hairy foreigners who had bossed the Vietnamese around for over half a century? And do we learn more about the reasons for the war from this figure of hundreds of thousands, than we do from the ratio 52 out of 54, which at one stage of the war was the proportion of South Vietnamese generals who had previously fought for the French against their own people?

Clearly no mathematical method, and no computer, can alone solve these dilemmas. A heavy dose of humanistic and moralistic training is necessary to evaluate them. Though it had many computer specialists, the United States went into Vietnam with few government employees who had undergone substantial training in Vietnamese language and culture. As late as the Chinese New Year's Eve ('Tet') in early 1968 — more than three years after substantial escalation of the American presence — we knew so little about Vietnamese history that American and South Vietnamese troops were dealt a severe setback by a surprise attack during holiday festivities, which amounted to a very close copy of one of the most famous battles

in Vietnamese history: a surprise infiltration and victorious battle led
by Nguyen Hui over invading Chinese during Tet in 1789 (Helen Lamb,
1972, p.58). Such was the result of the folly of imagining that
Communism was more important than Vietnamese nationalism in this
conflict.

It will be appreciated, from the gravity of these examples, that even
when one does not possess data or methods to achieve full mathematical
precision in answering large questions, well-informed people should make
the attempt anyway; or else ill-informed men, including politicians and
charlatans of all stripes, will make them instead.

Although the path to the great truths is often tortuous, the results
of the search may be clear and simple. Kepler's life and work illustrate
this as well as any in the history of science. He went up many blind
alleys and followed many mystical inspirations before coming up with
what are probably the most completely original of all the really
essential ideas in the history of science: the laws of planetary motion.
Yet these laws themselves are clear and simple. The same can be said
for Newton's law of motion and his law of gravity, for Darwin's
theory of natural selection and for most other great achievements in
science. The answers to the Vietnamese dilemma were not all that
difficult either: they involved principles like letting Congress declare
war, not lying to the American people and the simple but enormously
meaningful precept 'Thou shalt not kill'. The chain of rationalisations
to cover the arrogance, greed, fear and ignorance of the leadership is
what was too complicated.

It should be added that the situation in Vietnam could have been
better understood if environmental influences were taken more
seriously. These might have shown why a democracy could not be
reasonably expected to flourish there, why it was pointless to try
to create one by force and why our way of carrying out foreign policy
did not take into account the adjustments to the local environment
inherent in the lifestyles of the people. Environmental influences aside,
an appreciation of cultural differences alone would have helped a great
deal, of course.

*(5) Subjective truths are valued more highly, but objective truths are
usually more powerful in the long run.* Although I have implied that
many problems amenable to precise and quantitative solution are
unimportant, in fact they often seem important to us. We can measure
the size of a lot and measure the pieces which go to make a house

on the lot. These things are important to the builder, but chiefly in a subjective sense; they make little impact on the world as a whole. A flood or a hurricane or a glacier may destroy the house; it will be hard to predict any of these disasters, and even harder to control them, but they are more powerful than our human effort in building the house. Our own little worlds are full of these small facts which matter to us a great deal.

At times they are not facts, but merely subjective impressions. Our own subjective impression of space is important because it deals with the portion of space we live in. Still, it is also important to modify the errors in this impression. We may have an impression that a given floodplain is safe to build on. But then a flood may come in a few years and destroy our house. No matter how much we may value our subjective impressions in such cases we clearly would benefit by objectifying them in line with comprehensive knowledge of the world around us. It is here, in the effort to bring subjective reality into line with objective reality, that geographic perception studies could perform the greatest service; but too often they remain satisfied with observing how each person views the world differently, and often imply that these different impressions are equally valid. 'Valued', maybe; but 'valid', never. Science can have only one answer to a given question, or it is not science, not objective knowledge of any kind.

The search for objectivity beyond our private little worlds becomes more difficult as the problems become larger, as already suggested. It may even be very painful out on the 'edge of objectivity', as Gillespie has called it in his very powerful yet subtle analysis of leading trends in the history of science. Without some companionship in the search, it may be impossible. Solitary crusades like that of Kepler have been very rare. Newton could share coffeehouse conversation with other leading minds of his day over the nature of gravitational attraction; even at that, it cost him a nervous breakdown to formulate his results for publication in the *Principia*. Darwin also shared much conversation and correspondence over his work, less remote from the comfortable sphere of human life than Newton's; but he too suffered shattered health in the process of bringing it into presentable form. In the social sciences, we can also expect, despite the less precise form of most of our work, problems so difficult that they may be very painful to research and solve. But that is no reason to avoid them, or to rest content with subjective impressions far short of the edge of objectivity. It is instead reason to summon more forces and funds and personnel to bear upon the larger tasks. The main thing is to struggle to see beyond narrow personal limits. Life cannot consist only of this struggle, but it is the very essence of science and scholarship.

The distinction between the objective and the subjective may help explain the extraordinary insight into the concept of determinism made by philosopher Brand Blanshard. He pointed out that when planning for the future, we often seem to have many possible avenues we can follow; it seems merely a matter of choice. But when looking back later over the same situation we understand much better how our apparent free will was in reality constrained and conditioned to make the choice we actually made. Perhaps our current personal involvement in choosing paths leading from the present lends a subjective confusion to the situation, whereas looking back on the past, which in the nature of things can be more fully known than the future, constitutes more nearly an objective scientific evaluation, and that is why we can see the causes more clearly then.

None of this means that our individual free wills are totally without causal power; but that we are buffeted about by many forces stronger than ourselves, and what we count for as individuals is significant mainly on a subjective scale of values, more than in an objective calculation of forces.

(6) The environment is not stable, and therefore changing culture does not in itself prove lack of environmental influence. French possibilists in particular have argued that if culture changes it must be independent of the environment, which is stable. It comes down through the years from Vidal de la Blache and Brunhes to the more recent work of Perpillou (1966, p.48). But all of these possibilist ideas, from whatever year, remain logically prior to the thought of Huntington, who showed that climatic change during history was real and influential. Since 3000 BC, the atmosphere has been greatly changeable on scales important to human life; the new outpouring of climatic change studies since 1972 re-emphasises this circumstance many times over. And it also lends new strengths to theories of astronomical causes of climatic change, which is the type of theory Huntington advocated (see Huntington, 1923; Chappell, 1977. The definitive survey of knowledge about climatic change is Hubert Lamb, 1972-7).

In fact I believe that a logical extension of Huntington's thinking would tell us that the great periods of cultural creativity in our tradition themselves were largely determined by the cool, moist and variable weather of those times and by comparison with drier weather at other times. The periods involved would be principally that of the

Little Ice Age, about AD 1550 to 1850, when modern science and philosophy enjoyed their most fundamentally creative years; and the early part of the first millennium BC, extending somewhat into the latter half of that millennium, when Greek (pre-Socratic, Hellenic and Hellenistic all) and Hebrew culture, not to mention that of India and China, underwent remarkable and in a sense unrivaled development. What remains to be said below about the influence of climate on our thought and activity will possibly lend more credence to this claim.

(7) Cultural diffusion is not the logical antithesis of environmental causation. Both ideas can live in perfect harmony. In fact Huntington speculated on pre-historic trans-oceanic transmission of important cultural concepts, such as that of the Zodiac from Mesopotamia to Mexico (Huntington, 1919); and in no way did this vitiate his theories of the environmental influences on civilisation.

Yet many geographers have sought to deny environmental influences by showing that similar environments, say the Colorado and Nile valleys, produce different cultural responses, and that new ideas are so rare that they are re-discovered by someone else living in a similar environment. These ideas suffer from several flaws; for example: (a) No two environments are exactly alike, least of all the Colorado and Nile valleys, since only the Nile has regular natural floods which can guarantee agricultural prosperity. In other words, the earth is not uniform over space, any more than it is over time; (b) stage of cultural evolution, largely determined by density of population, must be a key factor in such comparisons, which no sensible environmental determinist would deny; and (c) multiple inventions have been very common in those cases in the history of science and technology for which we possess rather complete records. Even much of the work of Galileo and Newton was done prior to them in some other form, or simultaneously and independently elsewhere. A full dozen scientists share the glory for producing the laws of thermo-dynamics in the early nineteenth century. And there were many dozens of pre-Darwinian evolutionists, some of whom had the mechanism of natural selection in hand as well as the basic concept of organic change. Cultural geographers often concentrate on examples of inventions in primitive cultures about which little can be known with certainty; they might do better to dwell more often on the recent history of science in which the circumstances of invention are often known in great detail.

The greatest breakthrough in our knowledge of pre-Columbian

voyaging, in many years, perhaps ever, has come through the work of biologist and linguist H. Barry Fell (1976). He has discovered that 'ancient Maori', found on inscriptions in many parts of the Pacific realm, is actually a dialect from North Africa, where the names 'Mauri' and 'Mauretania' are related. A voyage from Libya set out in 234 BC to circumnavigate the world in order to prove correct the speculations of Eratosthenes, whose familiar proof of the size of the earth was carved on the rock walls of a cave in New Guinea by these voyagers. After this voyage reached the coast of Chile, it evidently turned back, and helped to populate and bring the art of long-distance navigation to Polynesia. Fell's remarkable work of decipherment and translation has also revealed the existence of Iberian Celts in North America as early as 800 BC. I mention these astounding findings partly out of enthusiasm to communicate them, while extreme sceptics in the archaeology and linguistics profession in the United States have in many cases spurned them, and partly merely to point out that they have nothing at all to say that is contradictory to environmental influences. In fact Huntington might try to interpret the story of the voyage across the Pacific from North Africa in terms of tracing the origin of the knowledge of long-distance navigation in the Pacific area to the more mentally stimulating cyclonic storm belt, rather than placing it in the monotonous equatorial region. And he just might have something; further work in reinterpreting Polynesian culture history in the light of Fell's findings will tell whether such an idea holds or not.

(8) Environmentally-shaped traits generated in one place may be carried to another place, where they will not be re-shaped immediately. This principle helps to explain why areas are not uniform culturally, and why such lack of uniformity does not necessarily prove that culture is independent of environmental influences. Another way of stating the same idea is to say it *takes time* for new arrivals in an area to become acclimatised, just as it takes time for vegetation, soils, or landforms to adjust to their environment.

Migrants from a culture hearth who take their environmentally-shaped culture with them into different environments can establish striking contrasts and anomalies. These would include, for example, Muslims who have reached to or into the humid tropics in Africa and Southeast Asia; European colonists who have built empires far from the seasonal and variable climates of their homelands; and Anglo-Americans who have pushed into the arid southwestern United States.

In many such cases, the boundary between the expanded culture area and the next neighbouring culture area offers such a sharp and obvious contrast that cultural determinists will cite it, noting that the physical environment scarcely differs on either side of the boundary, to try to prove that culture is not influenced by this environment. On the contrary, the contrast proves mainly that human beings have the power to carry their cultures over space. As time goes on, the new environment will indeed modify the transplanted culture, reducing the differences across the border.

Thus the southwestern US is becoming culturally more like northern Mexico and less like the northeastern US, as time goes on. Not only are Mexican-Americans increasing proportionally, but the southwestern American culture is developing more accommodations to the heat, aridity, monotony and lack of seasonality which were not dominantly present in the culture hearths of Europe and the Northeast. Californian culture becomes less hurried, less complex and individualistic, and more geared to relaxation rather than industrious effort, as time goes on, and a higher and higher proportion of its practitioners come to be born and raised locally, rather than in the climatically more stimulating Northeast and North Central region. As a native of California who has often returned for up-dated impressions of the state, I have a strong impression that over the last 40 years there has developed an increasing polarity in its culture, notably its political culture, with timid *status quo* worship on the one hand and emotionally-intense searching for extreme novelty on the other, while carefully rational individualism noticeably wanes. Such polarity is more characteristic of the tropics than of the mid-latitudes; but of course California is only in the subtropics, and its winters are generally moist and variable, though its summers are not — hence the departure from the original culture will be limited.

Also, such changes need not necessarily be for the worse. To take a couple of specific points, the environmental adjustment that may some day develop in the southwest is likely to include such wise moves as stopping work for more than a mere lunch hour during hot afternoons (even more appropriate for Arizona); and ceasing to try to grow humid, mid-latitude grasses as lawns on everyone's property, where water is so scarce. Mexicans have already made these two adjustments and it seems likely southwestern Americans will make them eventually also. Beyond these kinds of specific adjustments, a more calm and reflective culture in general would serve as a palliative to the often too-hastily active culture of the Northeast.

Even when cultures are fully in harmony with environment, or nearly so, they often display internal variety, based largely on class and/or racial distinctions.

Still, just as among siblings in a family, these differences largely reflect different experiences, including different cultural environments. The environment is still important, but here it is the cultural environment that predominates; and it deserves careful study and well-considered modification. At the same time it does not negate the often less-obvious influences by which the natural environment shapes the entire culture.

(9) Environmental influences are not limited to material phases of life; they also affect our minds and our habits. Huntington understood this principle from the first, even before coming up with the supreme gem of his entire body of thought, his theory of the stimulating influence of variable cyclonic weather. He substantiated this theory with data from both classroom and factory (Huntington, 1924).

General support for Huntington's approach to environmental influences is found in a wide range of experimental and theoretical reports from the leading laboratories of environmental physiology today, such as the John B. Pierce Laboratory of Hygiene in New Haven, Connecticut (for example, Nevins *et al.*, 1975); the US Army Research Institute of Environmental Medicine in Natick, Massachusetts, especially its Ergonomics Laboratory (Goldman, 1973); the Institute for Environmental Research, Kansas State University (Rohles, 1975); the National Swedish Institute for Building Research Environmental Laboratory (Wyon, 1974); and still others elsewhere. Of course these researchers carry the field Huntington called 'physiological climatology' to more advanced and multi-faceted levels of achievement than were possible in his time (see also Folk, 1974; and Slonim, 1974). The emphasis now is on atmospheric conditions inducing comfort, rather than on those inducing energy and creativity; but the facts are roughly the same from either point of view. So is the fundamental postulate that, in line with Darwinian principles, mankind evolved in adaptation to certain atmospheric conditions, and only these are now suitable for him to live in.

Another important centre of research on climatic influences on human activity has been the Biometeorological Research Centre in Leiden, the Netherlands. From this base, Solco W. Tromp has issued a large variety of publications, often ranging beyond concern with laboratory experiments, to broad philosophical issues (Tromp, 1967).

Cultural determinists continue to spread the myth that there is no

specific response to atmospheric variation, at least within a wide range.
Environmental physiologists also may claim that there is a range within
which responses are distinguishable; but the range is not all that wide,
and it is accompanied by realistic awareness of such involuntary
responses to atmospheric variables as sweating, spreading the limbs to
lose heat and shivering.

Huntington's ideas about cultural responses
to the atmosphere have solid roots in such physiological realities;
shivering, for example, is an extreme stage of cold-induced activity, the
purpose of which physiologically is to create more body heat by added
motion.

One contemporary physiologist, Auliciems, has produced a striking
study of the achievements of English school-children, which supports
Huntington's conclusions in many particulars: the children seem to
think more capably in cool, moist weather during the passage of a cold
front, when the temperature has just dropped slightly (Auliciems,
1972; review by Chappell, 1975).

Huntington also explored the influences of unusual atmospheric
phenomena such as ozone and air ions; in this work too he
foreshadowed later, more significant, developments in the same area
(McGregor, 1972; Krueger and Reed, 1976). Medicine, criminology
and architecture are fields which have been closely tied to theories
of climatic influences for centuries. Huntington made contributions
of some note in at least the first two of these areas, and received
an award for one of his medical articles (1921).

Huntington's work in physiological climatology has been criticised
because it comprises 'only correlations' rather than experiments.
Such antagonism overlooks the fact that he was careful to use sets
of data that would yield objective results, and that by examining work
records of factory piece workers after the work had been performed,
he eliminated the frequent problem, for testers of psychological
responses, of the 'Hawthorne Effect', i.e., the tendency to perform
differently than usually just because one is being tested.

It should be pointed out that there is such a thing as too much
cool and cloudy weather for maximal achievement. Yet even extremes
of this generally stimulating condition can have a special kind of
creative influence. During the exceptionally cool summer of 1816,
a group of important English literary figures, including Byron and
Shelley, were vacationing on the shores of Lake Geneva. Finding little
pleasure in nature in such circumstances, they decided to retire to
their desks and write ghost stories. From this effort came Mary
Shelley's *Frankenstein*; and *The Vampire* by John Polidori (We are

reminded of this by the French historian of climate, E. LeRoy Ladurie, who nevertheless maintains a generally sceptical attitude towards climatic influences on culture). June of 1816 produced snow in North America; it was the frostiest summer in American history. And its influence was not only on the mind and the spirit. Historian John Post blames the bad weather of this and surrounding years for *The Last Great Subsistence Crisis in the Western World*.

(10) Environmental determinism is not a politically reactionary philosophy, and it can be applied towards the solution of large practical problems. It has already been suggested that an awareness of environmental influences might have helped to avoid the pitfall of American involvement in Vietnam. But there is so much more to say in order to counteract the commonly-expressed belief that something reactionary and impractical inheres in the idea of environmental influences, that I will now develop the subject in a special section, with emphasis on problems of the Third World. Let me advise the reader at the outset that my goal in making these applications to political problems is not to move to the far left wing while escaping the dangers of far-right views. Rather I seek a sensible middle ground. If my views seem too far to the left for some readers, I would suggest to them that what is left-wing in the context of today's American politics is usually on middle ground in the context of world politics in general — and perhaps in the context of all of American history as well.

Progressive Applications to Contemporary Problems

The clearest charges of political reaction against environmental causation have come from Marxists of one variety or another; but it is worth noting that as the cultural thaw initiated by Nikita Khruschev developed, Soviet Russian geographers soon realised that not only were these charges oversimplified, but their own philosophy of environmental influences needed to be modified in terms of greater appreciation of environmental constraints (Chappell, 1975). They still retain great faith in technology, perhaps greater than American or Western European faith, but they have also learned the importance of ecological interactions, and have begun to practice considerable caution and restraint in such areas as water resource planning.

    Marxists and many others who place great faith in the ability of technology to increase food supplies continue to cast doubt on the Malthusian formula of geometric increase in food supplies. Indeed, since 1798 the human race has added to the food supply more substantially than at an arithmetic rate. But the vast expansion of cultivated grasslands has reached close to its necessary conclusion and increases of yields on currently ploughed lands may be reaching their limits too – at least, limits of available support technology such as irrigation water and fertilizer inputs, as many reviewers of the Green Revolution have been pointing out lately (Schneider, 1976; Roberts and Lansford, 1979). So perhaps the Malthusian environmental bind has only been delayed, as the Club of Rome reports now strongly suggest.

Let me emphasise that Malthus's largest, most obvious error was not his belief in the limitations of the earth's resources, which comprises a form of environmental causation theory; but rather his politically-motivated analysis of what makes population grow rapidly. He thought it happened as a result of giving too much food to the poor. But within a few years – as Hardin reminds us in his excellent compendium – observations by Doubleday and others were made to the effect that the fastest-growing populations were not those that were well-fed, but those that were ill-fed. The idea was slow to catch on. It was labelled 'revolutionary' when de Castro elaborated it in 1952, and only in the 1960s did it become a generally accepted part of 'demographic transition theory'. It is still debated, this tendency for material well-being and education to reduce birth rates; it may not apply to the Third World as clearly as it has to the developed world.

And yet it has plenty of evidence to back it up. At any rate, it is here, in Malthus's suggested solution to the problem, in his suggestion that we ought to keep the poor rather hungry so as to avoid overpopulation, that we must look for the truly reactionary element in his thought. In his appreciation of environmental influences we find nothing much worse than a delayed relevance.

None of this is to deny the great importance of the problems of land reform and economic imperialism, which, as pointed out with unprecedented thoroughness by Lappé and Collins (1977), lead to much or even most of the good land in the Third World being kept idle or being used to feed the wealthier nations. Yet we cannot solve the problems of hunger in the Third World by redistribution of resources only. The leached soils, the frequent extremes of weather, the often oppressive heat and humidity, the lack of stimulating change, the excessive carbohydrates as opposed to protein in crops that grow in wet climates, and other natural limitations will remain, and must be dealt with frankly. And after all, might not climate have something to do even with the fact that certain regions of the world contain most of the exploiters, and others more of the relatively passive victims of exploitation?

There are unrealistic extremes on the 'left' end of the political spectrum as well, and one of the most damaging to the progress of science was the episode of Lysenkoist genetics in Soviet Russia. But what did Lysenkoists affirm, other than that human will could overcome and speed up the processes of nature? Lysenko's error was not so much believing in Lamarckism, i.e., that characteristics acquired during one's lifetime can be passed on through the genes. For indeed there has been no conclusive refutation of Lamarckism; even Darwin leaned on it somewhat, and a few experimental results suggest at least vaguely that it might be active in nature. Rather, Lysenko erred in supposing that mankind itself could control and manipulate the process of inheritance of acquired characters, such as when vernalising wheat seed to produce a faster-growing strain. Darwinism, which argues that the environment produces new species very slowly by selecting the fittest, depends on environmental causation, whereas Lamarckism, and especially its Lysenkoist variant, depend on human will as causative agent. Of the two, the environmental-causationist approach has proven far more scientifically reliable in interpreting evolution.

On the reactionary side of the political spectrum we find social Darwinism, which in recent geographical literature, especially in America, has frequently been awkwardly misdefined as something akin to environmental determinism. True, the general notion of environmental causation receives rather considerable support from Darwin's work; but historically the term 'social Darwinism' has been applied to a narrower range of ideas within the Darwinian framework. Strict or crude social Darwinism is a philosophy of *laissez-faire* economics, advanced mainly by Herbert Spencer in England and by William Graham Summer in the United States, and based on the concept of 'survival of the fittest'. It argues that allowing free competition not only causes the strongest and the cleverest to dominate, but also is good for society as a whole. The term 'social Darwinism' has also been applied to answers to this crude faith in unrestrained competition, such as those by Kropotkin and Sheldon, who stress the importance of co-operation in survival (see Hofstadter, 1955). But I am using it in this chapter exclusively to apply to the original or crude variety of this concept.

Social Darwinist ideas have of course cropped up frequently in American politics without being called by that name. Barry Goldwater's platform in 1964 amounted to a wild-west version of social Darwinism. President Ford's refusal to give extra grain to the world's hungry at the World Food Conference in 1974 was another example. So was Daniel Moynihan's expression of 'benign neglect' as a statement of domestic welfare policy, during the Nixon administration. The point to be emphasised especially here is not so much the poverty of humanitarian outlook in these ideas and policies, which is obvious enough and of great importance, but rather the fact that they represent a belief outside of, and unconcerned with, the relation of mankind to his environment. The concern of 'social Darwinism' in its true historical sense, as opposed to its bogus adaptation to the campaign to discredit environmental influences, is almost exclusively with struggles among individuals, and not of individuals with environment.

Throughout recent American politics, one can find numerous examples of right-wing politicians, including social Darwinists insensitive to the needs of the underprivileged, talking within an artificial framework defined by both supporters and critics as 'the system', virtually unaware of environmental limitations either at home or abroad, and generally working to enhance large-scale private profiteering by the privileged few. In all of this, social Darwinism has been associated with culture and cultural causes, rather than with nature and natural-environmental causes.

Who has tried to turn us away from such unfortunate extremes? To some extent, until holding the power of high national administration distorted his view, Hubert Humphrey did. To a greater extent, never threatened by the opportunity to assume such great power, George McGovern did. Both men came from a South Dakota so marked by the human suffering of the Dust Bowl (the driest period in American history, the tree rings tell us, even if ploughing the land did make its results worse) that they could not help but appreciate the limits the environment puts on human efforts. McGovern talked often of the greatness of America's environment. His favourite campaign song was 'This Land is Your Land, This Land is My Land'. He saw through the artificialities of praising the 'system' as the rationale for fighting in Vietnam. He was and is a humane and progressive politician, and also one who does not live and think exclusively within culture, but often relates culture to environment (Anson, 1972; esp. p.26, citing McGovern).

These comments are not merely superficial appendages to recent political history. They have much to do with the most serious problems the American nation faces. I would even suggest that the biggest lie current in American life today is the lie which attributes the strength of the nation primarily to the 'system' of capitalistic free enterprise and democratic representative government. The truth, I suggest, is much closer to the following hypothesis; that the abundant American environment is the main cause of the democratic, free-enterprise system itself, as well as of the strength associated with it.

The correlation of democracy and competition with resource abundance is very strong throughout the world today and back into history. We can observe the correlation currently as energy resources become scarcer, and governmental controls, limiting freedom and competition, usually become tighter as a result. Most of the world is more collectivist and less democratic than America and Western Europe, because resources in most of the world are scarcer – and one should include the stimulating climate among the most valuable resources. A second political party, in fact, might be viewed as a luxury, a superfluous effort, where resources are scarce – resources for all phases of human activity, including agriculture, manufacturing and even the work of mounting political campaigns. This is not to *justify* the absence of democracy, but to explain it – the two are much different processes.

Those who argue that 'systems' are independent of environmental influences will also maintain that what is wrong with a country may be the result of its system. This assumption has often been used to criticise the Soviet Union. It is argued that all or most of their agricultural problems, for example, are due to shortcomings in their dictatorial, centralised economy, much more so than to those in their environment. At the 1977 national meeting of the Association of American Geographers, I heard one speaker carry this position to such a ridiculous extreme that he even claimed there were no famines in Russia before 1917; 'nature creates droughts', he asserted, 'while only man creates famines'. Of course mankind can create a famine, but to say that nature cannot is a horrendous distortion. Russian history tells of many famines before 1917, including several during and near the infamous 'Time of Troubles' in the early seventeenth century, which now can be well correlated with the bitter weather of the Little Ice Age.

We desperately need to replace such dangerous pseudo-science with realistic evaluations of environmental influences. A fine example is the comparison offered by Canadian meteorologist Derek Winstanley, who calculated that whereas 60 per cent of US agricultural land receives at least 28 inches of precipitation annually, only 1.1 per cent (yes, that is just *one point one* per cent) of Soviet agricultural land receives this same amount, an amount that allows the farmer to be comfortable in planting corn and soybeans. This assumes, of course, that the growing season is long enough and warm enough, which is another point on which nature has favoured the US much more than the Soviet Union (See Schneider, 1976, p.113).

Excess faith in cultural causation emanates from the Soviet camp as well. Schneider cites estimates of the number of people the earth can support comfortably, coming from both Soviet scholar Nikolai Rodionov and American scholar Roger Revelle (a publicly-declared fee of all kinds of socialism), and both men reach nearly the same conclusion: about 40 billion people (Schneider, 1976, pp.255-6). Most environmentalists are not so sanguine. At any rate, faith in technological power and cultural determinism clearly crosses ideological boundaries, and the key to realism will not be attached to any particular ideology, as these ideologies parade today on the world stage. Wisdom from all of them, reformulated and combined with much that has been forgotten or ignored, will be required to meet the crisis of the declining ratio of resources to population.

Among radical geographers and others leaning in the Marxist direction one occasionally hears the name of Karl Wittfogel as a Marxist who wrote like an environmental determinist and who also veered sharply away from Bolshevik Marxism — so much so that he supported social-Darwinist politicians in post-Second World War America. Is environmental determinism therefore to be correlated with reactionary politics? The life story of Wittfogel does not lend credence to this association. Wittfogel's appreciation of environmental influences in which he strained to achieve a balance between environmental and human forces met with disfavour on the part of Marxist editors. One of them published his 1929 article with a specific note disassociating the editorship from his ideas — which clearly were not far enough in the voluntarist or human-causation direction to be in step with what was happening in the Soviet Union at that time. And the follow-up article had to be published elsewhere in 1932 (see Chappell, 1967). It was in these articles that Wittfogel laid the groundwork for his interpretation of 'Oriental Despotism' as resulting from development of large-scale irrigation systems in arid floodplain environments. There is clearly a possibilistic element in this concept, and yet also a strong sense of environmental influence, which did not disappear just because Wittfogel inserted a small disclaimer in favour of human causation, in his *magnum opus* of 1957. In any event, it is clear from Martin Jay's biographical account that Wittfogel did not veer far away from Communism until well after 1932, after he had moved to America; and even then he adhered to a Marxism of a non-Bolshevik sort (Jay, 1973). The general tendency during his life seems to have been for his politics to move in the reactionary direction as his emphasis on environmental causation diminished.

Conclusion

In medieval times most of our ancestors believed that they lived at the centre of the universe. They also believed that only human souls could be immortal, whereas the earth and other living things had a finite existence. Today we congratulate ourselves that we have long since cast aside such geocentrism and subjectivity, and we credit ourselves with a more objective view of our places in space and time.

In the nineteenth century, disdain for 'uncivilised' people of the Tropics infused the writings of most European scholars, even those of Darwin. More recently, we like to boast, we have submerged this ethnocentrism and racism under a layer of more objective, scientific evaluations.

But we still have not cast off the conceit that culture in general is determined within itself and within the minds of those who practice it. A new kind of objectivity is needed related perhaps to religiously-based humility, which in its own way might lead to new scientific insights as important as those which followed from the writings of Copernicus at the dawn of modern science. This is the objective appreciation of environmental influences that should result from the logical extension of current ecological awareness, as discussed early in this chapter. Such an application of religious influence to science would be a far more sensible way to bring together these two areas of thought than is Lynn White, Jr.'s ill-founded attempt to blame Christian theology for ecological damage (see Barbour, 1973, for the pro and con of this argument). Religious values can teach the human race humility in the face of powers greater than itself, manifested on earth largely as laws of environmental influence over human life.

10 MARXISM: DIALECTICAL MATERIALISM, SOCIAL FORMATION AND THE GEOGRAPHIC RELATIONS

Richard J. Peet and James V. Lyons

Introduction

Any forthright consideration of the social theory of Karl Marx generally inspires discomfort if not outright contention. While such contention is, in many cases, based on some specific misunderstanding, it is both too frequent and too widespread to view simply as a problem of interpretation. In fact, one could conclude that Marx's analysis of capitalist production poses questions which should 'not be asked' and provides answers which should 'not be given'. In this respect, it is not only profoundly critical but genuinely threatening as well.

The Marxist perspective in American geography is of recent origin and in no small way evolved as a response to the developing material contradictions of American capitalism in recent years (Peet, 1977). Marx's method has proved particularly useful not only in relation to the traditional questions and problem areas within geography — human-environment relations, urbanism, regional development — but also as a theory of the history and development of geography itself as a concrete social activity. Although we will concentrate on basic concepts and topical applications, there is little doubt that the weakness of positive geographic theory cannot be viewed in isolation from its ideological/technical function in capitalist society. As one of many specialised work processes by which the existing social order reproduces itself, geography reflects its contradictions and exhibits its irrationality.

Dialectics

Although history had been studied for a long time before Marx, he was the first to arrive at a scientific understanding of history. Far from questioning the scientific status of history on the basis of its lack of regularity, Marx's social science would not only allow novelty but expect it. He drew from his study of political economy a profound realisation of new social forms and of the struggles from which these forms derived. He borrowed the dialectical method of Hegel as the best method for understanding these forms and the processes of their emergence and decline.

For Marx dialectics are not, as they were for Hegel, a creation of pure thought which is superimposed on nature and society and must presuppose their unreality. It is rather a method which evolved out of our active participation in historical process and our self-reflective desire to understand such participation. Accordingly, Marx took great pains to differentiate his use of dialectics from that of Hegel.

> My dialectic method is not only different from the Hegelian, but is its direct opposite. To Hegel, the life process of the human brain, i.e., the process of thinking, which, under the name of 'the Idea', he even transforms into an independent subject, is the demiurge of the real world, and the real world is only the external phenomenal form of 'the Idea'. With me, on the contrary, the 'ideal' is nothing less than the material world reflected by the human mind, and transformed into forms of thought (Marx, 1967, vol. I, p.19).

The dialectical method for Marx was therefore no substitute for empirical analysis and the gathering of facts but a way of dealing with their interrelations and dynamics. It means nothing more than viewing reality from the point of view of process and change.

Such change was for Marx internally generated by the ongoing dynamic of society itself. It takes the form of contradictions which prevent any system of production and attendant social relations from remaining in a state of equilibrium. Any readjustment or partial solution which attempts to mitigate the effects of societal contradiction induces new changes which open the way to new contradictions. As a result, society is in a constant state of flux.

For example, one of the principal contradictions of capitalist production for Marx was the tendency toward over production – the contradiction between production for need and production for profit.

> Since the aim of capital is not to minister to certain wants, but to produce profit, and since it accomplishes this purpose by methods which adapt the mass of production to the scale of production, not vice versa, a rift must continually ensue between the limited dimensions of consumption under capitalism and a production which forever tends to exceed this immanent barrier (Marx, 1967, vol. III, p.251).

As a result, certain mechanisms develop in capitalist society to alleviate the contradiction of overproduction — expansion of domestic and overseas markets, artificial depreciation of capital in the built environment and a territorial structure which promotes consumption, to mention a few. Although these mechanisms may prove successful in balancing productive capacity with social need in the short term, they cannot alter the basic tendency toward overproduction since it is inherent in the process whereby capital itself is accumulated by the capitalist class. These changes likewise help produce new contradictions — foreign conflict, urban decay, resource depletion, etc. — which in the end only serve to further enhance disequilibrium. These developments eventually lead to a state of crisis and potential breakdown. Thus:

> Crises are the real manifestation of the underlying contradictions within the capitalist process of accumulation. The argument which Marx puts forward throughout much of *Capital* is that there is always the potential within capitalism to achieve 'balanced growth' but that this potentiality can never be realized because of the structure of the social relations prevailing in a capitalist society. This structure leads individual capitalists to produce results collectively which are antagonistic to their own class interest and leads them also to inflict an insupportable violence upon the working class which is bound to elicit its own response in the field of overt class struggle (Harvey, 1978, p.11).

Materialism

Marx condemns the tendency of conventional social science to abstract from the real conditions of existence in capitalist society. He considers the monopoly of the capitalist class on the means of production of commodities, and therefore on the means of reproduction of life, to be a fact which impinges on all other aspects of social development. He does not conceive, however, that this arrangement arises out of some natural or eternal condition but rather that it is historically specific and socially defined by the power and property relations which are themselves in a constant state of flux. In this respect:

> Nature does not produce on the one side owners of money or commodities, and on the other men possessing nothing but their own labor-power. This relation is not one of natural history,

neither is it a social relation common to all historical periods. It is clearly the result of a past historical development, the product of many economic upheavals, of the extinction of a whole series of older forms of social production and the emergence of others (Marx, 1967, vol. I, p.169).

Such a view of history dismisses any attempt to represent social development as subject to immutable laws independent of historical process itself. Man is seen as an active social being in practice, transforming nature through labour for the purpose of material existence. Cognition, and therefore science, is seen as embedded in this process.

Marx begins with the premise of the existence of human individuals who must be in a position to live in order to be able to 'make history'. The first historical act is thus the production of the means to satisfy needs for food, drink, housing, clothing, etc., the production of material life itself. This 'mode of production' should not be considered merely as the reproduction of physical existence, but rather as a 'definite *mode of life*' (Marx and Engels, 1975, p.31). Individuals who are productively active in a definite way enter into definite social and political relations with one another; during their productive activity, humans also produce conceptions, ideas, etc. 'Consciousness can never be anything else than conscious being, and the being of men is their actual life process . . . It is not consciousness that determines life, but life that determines consciousness' (Marx and Engels, 1976, pp.36-7). Consciousness develops with productivity, the increase of needs, and of the numbers of people. It emerges as a dynamic process because in transforming the bounds of our experience, we transform ourselves. It develops especially with the division of labour, particularly the division between material and mental labour. From this point, consciousness may proceed to the formation of 'pure' theory, theology, philosophy, morality, etc. Hence the key to the understanding of the structure of social life and consciousness is the mode of production of the material basis of that life and consciousness.

This structure of understanding, which appears in a fully coherent form for the first time in *The German Ideology* (written between 1845 and 1847) was developed during the 1850s and appears in a mature form in Marx's Introduction to his *Critique of Political Economy*. We shall quote extensively from this Introduction, for it forms the basis of much of our subsequent discussion:

In the social production of their existence, men inevitably enter
into definite relations, which are independent of their will, namely
relations of production appropriate to a given stage in the
development of their material forces of production. The totality
of these relations of production constitutes the economic structure
of society, the real foundation, on which arises a legal and political
super-structure and to which correspond definite forms of social
consciousness. The mode of production of material life
conditions the general process of social, political and intellectual
life. It is not the consciousness of men that determines their
existence, but their social existence that determines their
consciousness. At a certain stage of development, the material
productive forces of society come into conflict with the existing
relations of production or – this merely expresses the same thing
in legal terms – with the property relations within the framework
of which they have operated hitherto. From forms of development
of the productive forces these relations turn into their fetters. Then
begins an era of social revolution. The changes in the economic
foundation lead sooner or later to the transformation of the whole
immense super-structure . . . In broad outline, the Asiatic, ancient,
feudal and modern bourgeois mode of production may be
designated as epochs marking progress in the economic development
of society. The bourgeois mode of production is the last
antagonistic form of the social process of production –
antagonistic not in the sense of individual antagonism but of an
antagonism that emanates from the individual's social conditions of
existence – but the productive forces developing within bourgeois
society create also the material conditions for a solution of this
antagonism. The prehistory of human society accordingly closes
with this social formation. (Marx, 1970, pp.20-2).

This passage interrelates mode of production with social formation
and presents a general theory of the change from one entire social
formation to the next. We shall analyse these essential components of
the dynamic Marxian theory of the structure of society in detail.

Mode of Production (the 'Economic Structure of Society')

The idea of mode of production as the foundation of the structure of
society is the logical conclusion of a materialistic philosophy. Under

this philosophy, of all possible relations between people, those
concerned with the production of the material basis of continued
life must be the most fundamental; these relations are structured
by the relations of production. And of all the physical infrastructure
and forces which make life possible, the forces of production are again
the most basic, the most fundamentally important.

The 'mode of production' is composed of two interrelated parts:
the relations and the forces of production. The *social relations of
production* are the forms of co-operation and mutual exchange of
activities necessary for production to take place; of particular
importance is the ownership of the means of production and thereby
the distribution of the social product (i.e., property relations). The
term *forces of production* refers to the technical way in which man
labours to transform nature into objects which have use value (utility).
'The elementary factors of the labour-process are: (1) the personal
activity of man, i.e., work itself, (2) the subject of that work and
(3) its instruments' (Marx, 1967, vol. I, p.178). 'Subject of work' means
raw materials, while the instruments of labour are tools, machinery
and infrastructure; together, the raw materials and instruments of
labour may be called the *means* of production.

For production to occur, labour must be combined with means
of production in a specific way; under the capitalist mode of production,
labour is separated from its independent means of production (during
'primitive accumulation') and forced to work with means owned by
the capitalist. Two consequences follow from this: 'First, the labourer
works under the control of the capitalist to whom his labour belongs . . .
Secondly, the product is the property of the capitalist and not that
of the labourer, its immediate producer' (Marx, 1967, vol. I, pp.184-5;
see also Balibar, 1970, pp.209-16). This product is pregnant with
surplus value which is the value of commodities produced by labour
power over and above the cost of reproducing the labour and means
of production used in any productive process. This surplus value is
realised when the product is sold, and flows back to the capitalist
where it forms the source of his property. Thus on the one side are
the workers who must sell their labour power in order to exist; on
the other side are capitalists who purchase labour power only to draw
surplus value from it. 'The existence of antagonist classes is thus
inscribed in production itself, *in the heart of production itself*: in the
relations of production' (Althusser, 1978, p.18). These antagonistic
relations of production pervade even the productive forces, since labour
power forms part of the productive forces, and since the process of

capitalist production always tends towards the maximum exploitation of labour power. Hence the technical mechanisms of production are subordinated to the class mechanisms of capitalist exploitation. Or, to put the same thing more generally, in the mode of production there is a unity between the productive forces and the relations of production *under the domination* of the relations of production.

By 'mode of production' therefore is meant a combination of forces and specific social relations of production and circulation of goods. It is assumed that to the mode of production (defined in a limited sense) there correspond various forms of political and ideological relations which are included in a wider conception of the mode of production (Godolier, 1977, p.18). No society continues in history unless it not only produces, but also reproduces, the material and social conditions of its existence. For capitalism, this means that the material, ideological and political conditions of exploitation must be reproduced *in production* (labour repression, anti-union activities, etc.) and *outside production* (by the state, the state ideological apparatuses, etc.). Interpreted in this way, Marx's *Capital* is no longer merely a theory of the political economy of capitalism but a theory of the material, legal-political and ideological forms of a mode of production founded on the exploitation of wage labour (Althusser, 1978, pp.19-20).

Social Formation ('Structure and Superstructure')

A given society is organised on the basis of several modes of production, some of which are in the process of disappearing, one of which is dominant and others of which are coming into being (see e.g. Godolier, 1977, pp.63-9). The concept of 'social formation' thus designates a social whole composed of distinct but interrelated 'instances' of the whole. Social formations are made up of 'levels' of these instances: those of its modes of production, or the economic structure — forces and relations of production — and those of the superstructures corresponding to these modes — politico-legal (law and the state) and cultural-ideological (religion, ethics, law, politics, etc.). The structure of relations between these levels has frequently been over-simplified into a fixed hierarchy in which the economic level *determines* in a simplistic, mechanical way the instances of the superstructure (e.g., Bukharin, 1925, ch.6). Yet, as the recent discussion centred around Althusser (1969; 1971; Althusser and Balibar, 1970) has shown, the interrelationship between the levels and instances of the social whole is highly complex.

That Marx himself was no simple economic determinist is shown by his statement:

> that the mode of production determines the character of the social, political, and intellectual life generally, all this is very true for our own times, in which material interests preponderate, but not for the middle ages, in which Catholicism, nor for Athens and Rome, where politics, reigned supreme ... This much however, is clear, that the middle ages could not live on Catholicism, nor the ancient world on politics. On the contrary, it is the mode in which they gained a livelihood that explains why here politics, and there Catholicism, played the chief part (Marx, 1967, vol. I, p.82n.).

In order to connect the 'two ends of the chain', on the one end determination in the last instance by the modes of production, on the other the relative autonomy of the superstructural instances, Althusser (1969, p.202) conceives the social formation as a structure articulated (combined) in dominance. In this structure the contradiction within the economic level, between the forces and relations of production (revealed in class antagonism) determines the character of the social totality because it determines which of the other instances is dominant. This version of structural causality is called 'over determination' by Althusser. When the mass of producers had their own independent means of production, surplus had to be extracted from them via the state or via ideology (e.g., religion), making either the political or the ideological instance of the formation dominant. Under capitalism, the producers (workers) have been divorced from ownership of their own means of production, surplus may be drained directly and the economic instance of the formation is both determinant and dominant. As Marx (1967, vol.III, p.791) puts it: 'It is always the direct relationship of the owners of the conditions of production to the direct producers ... which reveals the innermost secret, the hidden basis of the entire social structure' (see also Cutler, Hindess, Hirst and Hussain, 1977, p.177). In the Althusserian conception, each instance of the social formation moves through time with its own rhythm, unevenly developing relative to the other instances, with which it nevertheless is interrelated into an organic whole.

Such a complex theory of structural causality is necessary because the essential causes of things must be fetishised in the capitalist mode of production. Fetishism is the mode of existence of capitalist production, the very form which the system takes, and reality is not something underlying the appearances, but is the structured relations of these appearances (Callincos, 1976, pp.39-52). Althusser's interest is to explicate this structured relation. Whether he has entirely succeeded in doing so may be questioned (see e.g. Hindess and Hirst, 1977) but what we can learn from him is the need to understand the relations between the structure and superstructure in terms of dialectical determination, and the relative autonomy of the superstructure, rather than mechanical impress.

The same applies to the dynamics of social formations, the process by which contradiction builds into revolution. If the contradiction in the mode of production between the forces and relations of production is to become the source of 'revolutionary rupture' in the social formation, there must be an accumulation of 'circumstance' and 'currents', accumulation of contradictions, a fusion of contradictions from different origins. The general contradiction in the mode of production is present in all these circumstances, and even in their fusion, but we can no longer talk of its sole unique power. A revolution in the economic structure thus does not, *ipso facto*, modify the superstructural instances (and especially the ideological instance), for these have enough independence to survive a change in their immediate contexts, and even to recreate substitute conditions of existence. And the new society produced by revolution may itself ensure the survival of older elements through the forms of its new superstructure and specific (national and international) circumstances (Althusser, 1969, pp.98-106). Again, revolutionary change from one social formation to another has to be understood as a *dialectical* process.

What is the significance of this concept of social structure to geography? First, social formations have distinct interrelations with their natural environments. Second, the nature of spatial relations within and between social formations depend on the organising principles of these formations. In short, environmental relations (natural and spatial) are structured by the social formation, while the nature of the environmental relations into which social formations are embedded affects the development of these formations. Let us explore these in more detail.

Natural Environmental Relations

In the Marxist philosophy, man is a part of nature, immersed in it, yet also 'apart' as a conscious subject. Consciousness itself is natural in the sense that knowledge is drawn from nature via sense experience, and man's capacity for thought is a product of his nature.

First, let us look at the relation with natural environment in a most general way. In the materialist conception, the key interaction between man and nature is labour:

> Labour is, in the first place, a process in which both man and Nature participate, and in which man of his own accord starts, regulates and controls the material re-actions between himself and Nature. He opposes himself to Nature as one of her own forces, setting in motion arms and legs, head and hands, the natural forces of his body, in order to appropriate Nature's productions in a form adapted to his own wants. By thus acting on the external work and changing it, he at the same time changes his own nature. He develops his slumbering powers and compels them to act in obedience to his sway (Marx, 1967, vol. I, pp.177-8).

Man is confronted with a natural world which cannot be transcended and which must be appropriated in order to survive. His mode of appropriation is labour — yet labour itself is 'energy transferred to a human organism by means of nourishing matter' (Marx, 1967, vol. I, p.214, footnote no.1). Via a natural process, therefore, pieces of nature are transformed into use-values which serve as the material basis of further production and continued life. And through labour man also finds and makes his own 'human nature'.

In this dialectical relation with nature there are elements of 'determination in the last instance' reminiscent of the relation between the economic structure and the superstructure of the social formation. Labour, the 'metabolism' between man and nature, is 'an external nature-imposed necessity' (Marx, 1967, vol. I, pp.42-3) underlying all social formations. And the use of natural materials by humans is, in the end, only temporary as eventually they must sink back into the earth. An interruption of this process forms a contradiction which forces the restoration of the circulation of matter as a regulating law of social production (Marx, 1967, vol. I, pp.505-6). Hence 'the iron compulsion towards the production and reproduction of human life, which defines the whole of history, has in it something of the rigid cyclical form of nature' (Schmidt, 1971, p.90).

Characteristics of nature are imprinted on the histories of the social formation, but not in an obvious, mechanically deterministic way.

(See Burgess, 1978, for a Marxist critique of geographical determinism.)
The productiveness of labour is fettered by physical conditions. The
greater the natural wealth in means of subsistence (e.g., a fertile soil),
and in the instruments of labour (e.g., waterfalls, navigable rivers,
energy sources), the less the labour time necessary for the maintenance
and reproduction of the producer, thus the more labour for others
that can be performed. But it does not follow from this that simple
natural fertility is necessarily the most suitable environment for social
development.

It is not the mere fertility of the soil, but the differentiation
of the soil, the variety of its natural products, the changes of the
seasons, which form the physical basis for the social divisions of
labour, and which, by changes in the natural surroundings, spur
man on to the multiplication of his wants, his capabilities, his
means and modes of labour (Marx, 1967, vol. I, pp.513-14; also
Parsons, 1977).

Natural conditions alone give only the *possibility*, never the reality,
of surplus-labour and surplus value.
    Now let us look at natural environmental relations more specifically.
Labour is not only the fundamental human relation with nature, but
also the fundamental relation between people. Labour transforms
natural objects into use-values in the context of particular social
relations. Different modes of production have natural environmental
relations which reflect the character of their dominant social relations.

Thus under capitalism men struggle with nature in order to satisfy
new needs, but they do so in a prescribed way (namely under
conditions of wage-labor) that differs profoundly from other modes
of production . . . the antagonisms of a class-divided society makes
it impossible for men to bring their productive system (of which
mastery over nature is a part) under their control (Leiss, 1974,
pp.84-5).

Let us briefly apply this idea to an analysis of capitalism's present
relations with its natural environment, and especially to the onset of
environmental crisis in the late stages of capitalist mode of production.
    Two linked characteristics of the capitalist mode of production are
of particular interest. Firstly, the competitive nature of capitalism
compels constant economic expansion (i.e., the accumulation of capital):

The development of capitalist production makes it constantly
necessary to keep increasing the amount of capital laid out in a
given industrial undertaking, and competition makes the immanent
laws of capitalist production to be felt by each individual capitalist,
as external coercive laws. It compels him to keep constantly
extending his capital, in order to preserve it, but extend it he cannot,
except by means of progressive accumulation (Marx, 1967, vol. I,
p.592).

Accumulation, at the societal scale, leads to an expansion in the demand
for raw materials, with pressure being focused on certain crucial areas
such as energy (the demand for non-recyclable energy rises more
rapidly than accumulation).

Secondly, 'the directing motive, the end and aim of capitalist
production, is to extract the greatest possible amount of surplus-value,
and consequently to exploit labour-power to the greatest possible
extent' (Marx, 1967, vol. I, p.331). This characteristic of the economic
structure is reflected in, and reproduced by, the cultural instance in
the superstructure: an exploitative, competitive, alienated, commodityised
population is necessary to perpetuate an exploitative, competitive,
alienating mode of commodity production. These cultural characteristics
of social relations extend into the social formation's environmental
relations. Hence an economic system compelled to expand production
by its own inner laws, characterised by aggressive, exploitative social
and environmental relations, *necessarily* comes into a contradictory
relationship with a finite, fragile natural world (Peet, 1977; 1978a;
1979b). This contradiction, which is an era of high technology
threatens to destroy the natural basis of life, assumes a central role in
the stage of late capitalism; environmental crisis interacts with and
structures the effects of other contradictions, to bring about an
eventual revolutionary change to a new mode of life.

Spatial Relations

The materialist approach to space treats it as an integral part of a
general social theory: 'men . . . enter into particular social relations,
which give to space . . . a form, a function, a social signification'
(Castells, 1977, p.115). Space may be analysed as an 'expression' of
the social structure; this amounts to 'studying its shaping by elements
of the economic system, the political system and the ideological

system, and by their combinations and the social practices that derive from them' (Castells, 1977, p.126). Thus the economic structure is expressed spatially in the locations of production, consumption (reproduction of labour power) and exchange; the politico-legal instance is expressed, through its functions of domination-regulation and integration-repression, in the political segmentation of space and its actions on the economic organisations of space; and the ideological instance is expressed in cultural forms and symbolic meanings in landscapes. The social organisation of space is determined by each of the elements of the three instances (and by combinations of the elements of any one instance), by combinations of the three instances, by the persistence of spatial forms created by earlier social structures articulated with the new forms and by the particular actions of individuals and social groups on new environments. However, in the Althusserian conception of social formation, the economic level is both determinant and dominant under the capitalist mode of production. Hence for Castells (1977, p.130), writing under Althusser's influence, under capitalism 'the production element is the basis of the organisation of space'.

The total ensemble of the spatial effects of a mode of production has been termed 'territorial structure' by Buch-Hanson and Nielson (1977; see also Corragio, 1977), while for Santos (1977, p.3) it is impossible to speak of social formations without including the category of space; hence he suggests we re-name them social and spatial formations. But this discussion is too recent for a clear set of accepted terms to have evolved. And it will take some time for agreement to be reached, for the relations between social processes and space are the subject of an ongoing controversy within 'Marxist Geography'. This controversy concerns the degree of autonomy which space possesses; that is, to what degree is space 'a separate structure' with its own laws of inner transformation (Harvey, 1973, pp.302-13)? The idea of an autonomous space, associated with the work of Lefebvre (1972), has been attacked as 'spatial fetishism', as a diversion from the more fundamental issue of the social relations of production. However, recently, a milder version — the socio-spatial dialectic — has been advanced by Soja. Soja's view is essentially that:

> The basic structures and contradictions in the capitalist (and any other) mode of production are expressed simultaneously and dialectically in both social and spatial relations . . . not only does the organization of space express social relations but . . . social

relations (and hence class structure) are themselves, to an important degree, expressions of the spatial relations of production (Soja, 1978).

In this quotation, spatial relations are defined in terms of the organisation of space for the reproduction of labour power and the maintenance of the social order and for production. Space has a *relative* autonomy, like the instances of the social formation discussed previously. Hence, although considerable debate will undoubtedly follow, Soja's position is more nearly compatible with existing, accepted concepts of Marxist geography. In the meantime, however, the whole conception of the social formation as a totality generated by an organising principle, determination by the economy in the last instance, has been critiqued by Hindess and Hirst (1977) who have been hastily followed by at least one geographer (Gregory, 1978, pp.105-22). These ongoing fundamental discussions should eventually serve to strengthen and deepen the materialist conception of space and spatial relations.

All this has been very general and philosophical. To exemplify, let us give a brief example of Soja's socio-spatial dialectic, the structuring of world space by the capitalist social process and the conditioning of ongoing process by the resulting spatial structure. We shall begin by recalling that the essential characteristic of capitalist social relations is the transfer of value from the direct producers (workers) to the owners of the means, and controllers of the conditions, of production (Marx, 1967, vol. I). It is clear that neither production nor ownership occur in a spaceless vacuum – both have a location. Hence the main *social* relationship of capitalism is immediately its outstanding *spatial* relationship; the transfer of surplus value over space. This transfer may be: local, for example between workers based in one district of an industrial town and the bank of a factory owner living 'at the top'; inter-regional, for example between coal miners of Appalachia and the coal company owners of New York City; or at a world scale, for example between the plantation workers of India and the plantation owners of 'London and the Riviera', of 'Boston and Palm Beach', etc. In an immediate sense, this is not a transfer of value between regions, but between classes of people located in different regions, or in different areas of the same region. And the relation between regions is not a purely 'spatial' relation (what could that be?) but a socio-spatial relation. Even so, the *geography* of social relations is a highly significant aspect to these relations (see, e.g., Harvey, 1975). For, on the basis of an infusion of surplus value, some

regions come to have entire structures of social and economic
characteristics which are different from those of the regions from
which surplus value is being drained. These different structures
embedded into space may be called 'geographical specificities' of the
capitalist peripheral formations between which 'are relations of
domination, unequal relations expressed in a transfer of value from
the periphery to the center'. (For a discussion of the mechanisms
by which value is transferred see Emmanuel, 1972, and Szentes, 1971.)
And the socio-spatial relations which pass between these geographical
specificities come to be charged with a combination of characteristics:
not just classes interact across space, but class societies of one capitalist
type socio-spatially relate with class societies of another capitalist type.
The peripheries are thereby not only subjected to economic
exploitation as the basic socio-spatial relation, but also to the
undermining of their cultures and societies in manifold ways, as
additional elements to that relation. This total process of destructive
exploitation is 'underdevelopment'.

The implications of this concept are of enormous importance.
As Frank puts it:

> Underdevelopment is not due to the survival of archaic institutions
> and the existence of capital shortage in regions that have remained
> isolated from the stream of world history. On the contrary,
> underdevelopment was and still is generated by the very same
> historical process which also generated economic development:
> the development of capitalism itself (Frank, 1969, p.9).

Hence, for example, a proposal for a system of 'growth centres' and
'improved transport' in a Third World region made in the well-
intentioned belief that it will aid in economic development, also has
the result of making more efficient the drain of surplus value towards
the world capitalist centre. What Marxists call 'bourgeois geography'
(because in the end it acts in the interests of Capital) is thus in direct
conflict with Marxist geography. To summarise: a social relation
(exploitation) acts over space to create a geography of exploitation
which then acts as the context in which new versions (total
underdevelopment) of capitalist socio-spatial relations emerge.

Marxist Geography

Marxist science rests on, and is integrated by, its assumptions about the

importance of material production in structuring social processes. The
Marxist geography shares these assumptions: it is that part of a whole
science which specialises in the dialectical relations between social
processes on the one hand and the natural environment and spatial
relations on the other. Like the other Marxist sciences it is aimed at
changing the fundamental operation of social processes by changing
the social relations of production. Social revolutionary changes are
necessary to solve endemic spatial and environmental problems, for
these problems originate deep in the capitalist social formations.

This much should be clear from the preceding discussion. Let us
now take the argument a step further. A social formation moving
through time is made up of many particular social processes, but is
also a whole (general) process. This whole process interacts with
the existing natural and social environment. At one level this interaction
also produces geographically specific varieties of the social
environment; at another, it produces geographically specific varieties
of the social formation which is a localised version of the social process.
In each of these geographically-embedded pieces of the social
formation, the general process of historical change goes on somewhat
differently; that is, in different ways and at different speeds (Figure
10.1). The build-up of contradictions at one place varies in content

Figure 10.1: Social Processes and Spatial Relations

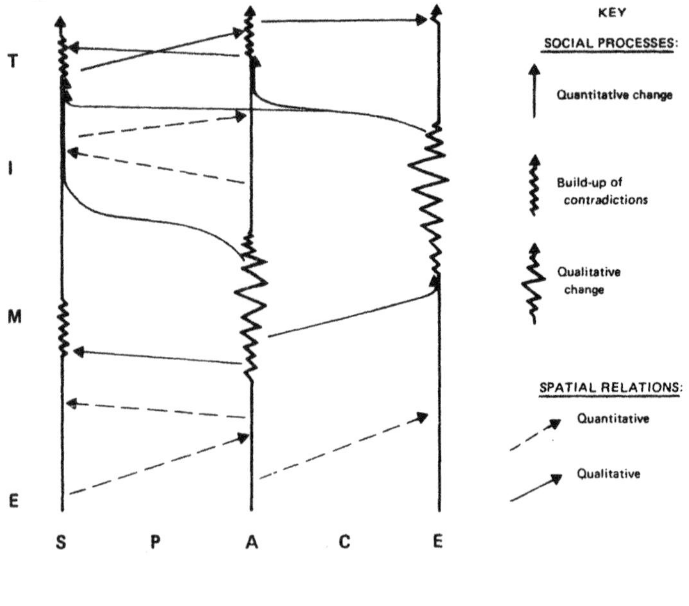

and tempo from those at other places. Yet as geographical components of a whole system, these geographical specificities are also related across space via socio-spatial relations. These socio-spatial relations not only reflect, or express, the content of the particular localised social processes which originate them, but also serve those processes in functional ways. As a localised process becomes enmeshed in contradiction in one area, it 'uses' its relations with processes in other areas to relieve the pressures which result from the build-up of its internal contradictions. 'Uses' is in quotation marks in the last sentence because it is a shorthand, personified term expressing an extremely complex relation between structural causality and the surface consequences and events. In relieving the pressure of the intensification and interaction of contradiction at one place, via socio-spatial relations with other places, capitalism injects more intense contradictions into the entire socio-spatial structure of its world system, which then rebound on the original source. Marxist geography, in its more sophisticated state, will be the study of the spatial relations between geographically specific versions of the general build-up of the inherent contradictions of capitalism. Let us conclude with an example which shows the operation of this system of geographically interrelated contradictions and, at the same time, integrates the two main components of environmental relations (natural and spatial relations) into one analysis.

The previous discussion on spatial relations divided what Wallerstein (1974) has shown to be a single world capitalist system into geographically specific versions: a world central social formation, towards which surplus value flows (the First World), and a series of peripheral formations, from which surplus value is drained (the Third World). Surplus value (capital) accumulated in the centre (from various sources, including local sources) is used to support economic production in the next round of 'development'. As the owners of capital produced by others, the business and banking institutions of the centre allocate and re-allocate forms and amounts of production to different regions of the world capitalist system by withdrawing capital from the regions where it has historically been produced by the local working class, and investing it elsewhere to breed more surplus value. Regional 'development' at one place is thus linked through a system of socio-spatial relations to regional underdevelopment elsewhere.

The particular form of exploitation of the Third World by the First changes over time. One way to classify these changes is in terms of the forms of capital flows between the two sets of social formations. Thus Palloix (1977) has divided the capital (and thereby social) relations between the First and Third Worlds into stages: the 'internationalisation' of commodity-capital, then money-capital and now productive capital. The last of these stages essentially corresponds to the international production of the multinational corporations. Thus, for example, by the early 1970s international production by US-controlled firms was worth $172 billion a year, which was four times greater than the value of US exports at that time (Palloix, 1977, p.8). Obviously there is a tremendous flow of capital out of the centre to support manufacturing and other forms of production, in the cheap-labour regions of the Third World (especially in 'free trade zones') as the prelude to even greater return flows of surplus value. What are these flows out of the centre a response to?

Capitalism is a system propelled through time through the development and interaction of its inherent contradictions. The longer it exists in any region the more intense and interlocked these contradictions tend to become, and the more drastic their social and environmental consequences. In the old centres of capitalist production two types of highly developed contradiction are most evident. First, contradiction between the forces and the social relations of production, revealed in class struggle largely of an economistic type, yields higher wages for the organised working class, the diversion of some surplus value away from the capitalist class and lower rates of profit. This contradiction is also reflected in social problems of various kinds which have to be contained and controlled by a state supported by high taxes, which constitute a further drain on surplus value and profit. Second, contradiction in the environmental relations of capitalist production, are revealed in shortages and high prices for raw materials, high direct and indirect costs from pollution and other damages to the natural world, hence lower rates of profit (Peet, 1979a). One of Capital's responses to its development of these contradictions is to abandon the old industrial regions of the centre (Northern England, New England, etc.) in search of virgin environments (to despoil), ideologically and politically virgin labour forces (to exploit), and higher profits. The internationalisation of production is thus the spatial response to the intense development of contradiction at the centre.

A new version of the world's capitalist system thus begins to take shape. Its central institutions are transnational corporations with many bases of power. Resistance to a corporation at one place is met by the withdrawal of capital from that place and its investment elsewhere.

The power of this new form of capitalism to control the world's people thus becomes greater than ever. If we want to understand what is happening to us, the people of the capitalist world, if we want to resist total control, we need a powerful, logical, sequential, political form of analysis to give an accurate, deep portrayal of the causes of events. Vague theories with unspecified political attachments will be taken over as protective ideologies by a virulent capitalism (Peet, 1978b). By contrast, Marxism and Marxist geography provide a powerful theoretical and political base for resistance. They are theories constructed on behalf of the mass of 'ordinary' people of the world to aid in our struggle against an international ruling class and a destructive and exploitative form of social life.

GLOSSARY

Anthropomorphism: The attribution of human qualities to non-human forms.

Axiom: A truth which is self-evident or a universally accepted principle or rule. Basic building block in theories and laws.

Concatenation: The process of linking or connecting a series of things or events together in a chain.

Deduction: A form of inference whereby a scientific system is arranged in such a way that from some of the premises and propositions all other propositions must follow logically.

Determinism: The view that all human behaviour is entirely controlled by some law or force which elicits a specific reaction.

Dialectic: A method of reasoning in which contradictions or opposites are synthesised and where thought proceeds from thesis to antithesis to synthesis.

Empiricism: The belief that knowledge is the result of experience.

Epistemology: That branch of philosophy which is concerned with the study of the nature of knowledge, the source of knowledge, and the criterion of knowledge.

Existentialism: A philosophical view that man is responsible for making his own nature. It stresses personal freedom, personal decision and personal commitment.

Humanism: A viewpoint that emphasises distinctively human ideas, interests and values. In geography it has been associated with philosophies like phenomenology and existentialism.

Hypothesis: An assumption, postulate or supposition which is as yet unproven but which is accepted tentatively as a basis for investigation.

Idealism: A view that reality is mental or mind-dependent.

Ideographic: A method which stresses the individuality and uniqueness of phenomena as opposed to the similarities phenomena have with each other.

Induction: A form of inference whereby conclusions about a class of phenomena are made on the basis of the observation and study of a set of specific cases.

Instrumentalism: An alternative name for Pragmatism.

Marxism: The system of thought developed by Karl Marx which claims that the state, through history, has been a device for the exploitation of the masses by a dominant class and that class struggle has been the main agent of historical change.

Materialism: The doctrine that reality is matter. Mind and consciousness are merely manifestations of such matter and are reducible to its physical elements.

Metaphysics: (beyond physics) A critical study of reality. It generally asks questions that relate to the nature of reality, the types of reality and the goals of reality.

Naturalism: A philosophical viewpoint that does not believe in a supernatural being. It contends that nature or the empirical world is the whole of reality.

Nominalism: A philosophy that believes that only particulars are real and that universals, or general terms, are only names.

Nomothetic: A method which is law-seeking and concerned with the establishment and verification of generalisations about phenomena.

Ontology: A branch of metaphysics (the study of reality) which is concerned with the nature of being as distinct from material existence.

Phenomenology: A philosophical school associated with Edmund Husserl (1859-1938) that underscored concern for what is perceived and the general patterns of consciousness and experience of individuals and groups.

Positivism: A philosophical viewpoint that limits knowledge to facts that can be observed and to the relationships between these facts.

Premise: A proposition which constitutes the basis for a conclusion.

Pragmatism: A philosophy which asserts that meaning and knowledge can only be defined in terms of their role in experience. It emphasises experiences, experimental inquiry and truth as criteria for evaluating consequences.

Rationalism: Reason is the source and the criterion of truth. This viewpoint believes that reason alone is a source of knowledge and is independent of experience.

Realism: A view that reality exists independent of the mind; it is not mind-dependent.

Relativism: A view that an individual's judgement is relative to place, time and value systems.

Scientific Method: The processes and steps by which a science obtains knowledge.

These normally include the identification of a problem, formulation of a hypothesis, gathering of relevant information against which to test the hypothesis, and the drawing of a conclusion with respect to the validity of the hypothesis.

Solipsism: The view that one (the individual) alone exists.

Teleology: A form of explanation that is based or focuses on the ends as a means of understanding the cause rather than on the causes that bring about the event.

Theorem: A theoretical proposition embodying something to be proved. Logically deduced from a set of axioms.

Theory: A more or less established or verified explanation accounting for known facts or phenomena. A set of propositions used as principles of explanation, not completely based on empirical evidence.

BIBLIOGRAPHY

Abler, R., Adams, J. and Gould, P. R. (1971) *Spatial Organization: The Geographer's View of the World* (Prentice Hall, Englewood Cliffs)

Achinstein, P. and Barker, S. F. (eds.) (1969) *The Legacy of Logical Positivism* (John Hopkins, Baltimore)

Ackerman, E. A. (1963) 'Where is a Research Frontier?' *Ann. Ass. Amer. Geogr.*, 53, 429-40

Acton, H. B. (1967) 'Idealism', in P. Edwards, (ed.) *The Encyclopedia of Philosophy* (Macmillan and Free Press, New York), IV, 110-18

Alexander, S. (1920) *Space Time and Deity*, 2 vols. (Macmillan, London)

Althusser, L. (1969) *For Marx* (Pantheon Books, New York)

———— (1971) *Lenin and Philosophy and Other Essays* (New Left Books, London)

———— (1978) 'Marxism-Leninism and the Class Struggle', *Theoretical Review*, 1, 17-20

Althusser, L. and Balibar, E. (1970) *Reading Capital* (New Left Books, London)

Amedeo, D. and Golledge, R. G. (1975) *An Introduction to Scientific Reasoning in Geography* (John Wiley, New York)

Amin, S. (1974) *Accumulation on a World Scale*, 2 vols. (Monthly Review Press, New York)

———— (1976) *Unequal Development* (Monthly Review Press, New York)

Anson, R. S. (1972) *McGovern: A Biography* (Holt, Rinehart and Winston, New York)

Auliciems, A. (1972) *The Atmospheric Environment: A Study of Comfort and Performance* (University of Toronto Dept. of Geography, Toronto. Research Publications no. 8) (Reviewed by John E. Chappell, Jr., in *Geographical Journal*, 141, 1975), 125-6

Aune, B. (1970) *Rationalism, Empiricism, and Pragmatism: An Introduction.* (Random House, New York)

Bachelard, G. (1969) *The Poetics of Space* (Beacon Press, Boston)

Balibar, E. (1970) 'On the Basic Concepts of Materialism', in Althusser and Balibar, *Reading Capital* (New Left Books, London), 201-308

Barbour, I. (ed.) (1973) *Western Man and Environmental Ethics: Attitudes. Towards Nature and Technology* (Addison-Wesley, Reading, Mass.)

Bartels, D. (1973) 'Between Theory and Metatheory' in Richard Chorley (ed.), *Directions in Geography* (Methuen, London), 25-37

Beaujeu-Garnier, J. (1974) 'Toward a New Equilibrium in France?' *Ann. Ass. Am. Geogr.*, 64, 113-23

Beck, R. N. (1969) *Perspectives in Philosophy* (Holt, Rinehart and Winston, New York)

Bergman, F. (1975) *Modern Political Geography* (WM C. Brown, Dubuque, Iowa),

Berthelot, R. (1949) *La Pensee de l'Asie et l'astrobiologie* (Payot, Paris)

Billinge, M. (1977) 'In Search of Negativism: Phenomenology in Historical Geography', *Journal of Historical Geography*, 3, 55-67

Blanshard, B. (1958) 'The Case for Determinism', in Sidney Hook (ed.), *Determinism and Freedom in the Age of Modern Science* (New York University Press, New York), 3-15

Bloor, D. (1975) 'A Philosophical Approach to Science', *Social Studies of Science*, 5, 507-17

Boorman, H. L. (1968) 'Mao Tse-Tung as Historian' in A.Feuerwerker (ed.), *History in Communist China* (MIT Press, Cambridge)

Brown, L. (1973) 'Diffusion Processes at the Macro and Meso Scales: An Update on a Framework for Future Analysis' (Ohio State University, Columbus).

Brown, R. (1968) *Explanations in Social Science* (Aldine, Chicago)

Brunhes, Jean (1920) *Human Geography* (Rand McNally, Chicago); translated by T. C. LeCompte and edited by Isaiah Bowman and R. E. Dodge; from second French edition, 1912

Buber, M. (1957) 'Distance and Relation', *Psychiatry*, 20, 97-104

Buch-Hanson, and Nielson, B. (1977) 'Marxist Geography and the Concept of Territorial Structure', *Antipode*, 9, 1-12

Bukharin, N. (1925) *Historical Materialism* (International Publishers, New York)

Bunge, W. (1962) *Theoretical Geography* (Gleerup, Lund, Sweden)

Bunting, T. E. and Guelke, L. (1979) 'Behavioral and Perception Geography: A Critical Appraisal', *Ann. Ass. Am. Geogr.*, 63, 448-62.

Burgess, R. (1978) 'The Concept of Nature in Geography and Marxism', *Antipode*, 10, 1-11

Burton, I. (1963) 'The Quantitative Revolution and Theoretical Geography', *The Canadian Geographer*, 7, 151-62

Buttimer, A. (1970) *Society and Milieu in the French Geographic Tradition* (Rand McNally, Chicago)

———— (1974) 'Values in Geography', *Resource Paper*, No. 24, Commission on College Geography, Association of American Geographers (Washington, DC)

———— (1976) 'Grasping the Dynamism of Lifeworld', *Ann. Ass. Am. Geogr.*, 66, 277-92

Callincos, A. (1976) *Althusser's Marxism* (Pluto Press, London)

Caro, R. (1975) *The Power Broker: Robert Moses and the Fall of New York* (Vintage Books, New York)

Castells, M. (1977) *The Urban Question* (Edward Arnold, London)

de Castro, J. (1952) *The Geography of Hunger* (Little, Brown, Boston)

Chappell, John E., Jr. (1967) 'Marxism and Environmentalism', *Ann. Ass. Am. Geogr.*, 57, 203-6

———— (1969) 'On Causation in Geographical Theory', *Proceedings Ass. Am. Geogr.*, 1, 34-8

———— (1971a) 'Climatic Pulsations in Inner Asia and Correlations Between Sunspots and Weather', *Palaeogeography, Palaeoclimatology, Palaeoecology*, 10, 177-97

———— (1971b) 'Huntington, Wittfogel, and the Environmental Bases of Ideas and Politics', *Selected Papers* of 21st International Geographical Congress (New Delhi, 1968), 3, 364-70

———— (1972) 'The Incomplete Chain of Causation: Barrier to Problem-Solving', *Journal of Geography*, 71, 452-6

———— (1974) 'Journal Interview', *Journal of Geography*, 73, 33-43.

—— (1975) 'The Ecological Dimension: Russian and American Views', *Ann. Ass. Amer. Geogr.*, 65, 144-62

—— (1977) 'Lake Levels and Astronomical Theories of Climatic Change', in Deon C. Greer (ed.), *Desertic Terminal Lakes* Ogden, Utah, May 1977; published by Utah Water Research Laboratory, Utah State University, Logan, Utah 84322, 27-35

Chodak, S. (1973) *Social Development* (Oxford University Press, New York)

Chorley, R. J. (1962) 'Geomorphology and General Systems Theory', in F. E. Dohrs and L. M. Sommers (eds.) *Introduction to Geography: Selected Readings* (Thomas Y. Crowell, New York)

Chorley, R. J. and Haggett, P. (eds.) (1967) *Models in Geography* (Methuen, London)

Clark, P. J. and Evans, F. C. (1954) 'Distance to Nearest Neighbor as a Measure of Spatial Relationships in Populations', *Ecology* 35, 445-53

Cohen, M. and Nagel, E. (1934) *An Introduction to Logic and Scientific Method* (Harcourt Brace, New York)

Collingwood, R. G. (1946 and 1956) *The Idea of History* (Oxford University Press, New York)

Collingwood, R. G. and Myers, J. N. L. (1937) *Roman Britain and the English Settlements* (Oxford University Press, London)

Commoner, B. (1971) *The Closing Circle* (Knopf, New York)

Corragio, J. (1977) 'Social Forms of Space Organization and Their Trends in Latin America', *Antipode,* 9, 14-28

Croce, B. (1941) *History as the Story of Liberty* (Henry Regnery Company, Chicago)

Cutler, A., Hindess, B., Hirst, P. and Hussain, A. (1977) *Marx's Capital and Capitalism Today*, vol. 1 (Routledge and Kegan Paul, London)

Dardel, E. (1952) *L'Homme et la Terre: Nature de la Realite Geographique* (Presses Universitaires de France, Paris)

Davidson, D. (1974) *Can Africa Survive: Arguments Against Growth Without Development* (Little, Brown, Boston)

Davis, K. (1967) 'The Myth of Functional Analysis as a Special Method in Sociology and Anthropology' in N. J. Demerath III and Richard A. Peterson (eds.) *System, Change and Conflict* (Free Press, New York), 379-402

Davis, W. M. (1906) 'An Inductive Study of the Content of Geography', *Bulletin of American Geographical Society*, vol. 38, 67-84 (Presidential address to second meeting of Ass. of Amer. Geogr., 1905; reprinted in W. M. Davis, *Geographical Essays*, 1909, 3-22)

Dickinson, R. E. (1969) *The Makers of Modern Geography* (Praeger, New York)

Dooley, P. K. (1974) *Pragmatism as Humanism: The Philosophy of William James* (Nelson-Hall, Chicago)

Durkheim, E. (1938) *The Rules of Sociological Method*, trans. Sarah A. Solovay *et al.* (Free Press, New York)

—— (1951) *Suicide*, trans. John A. Spaulding and George Simpson (Free Press, New York)

—— (1973) *The Division of Labor in Society*, trans. with introduction by George Simpson (Macmillan, New York)

Eiseley, L. (1958) *Darwin's Century* (Doubleday, Garden City, NY)

Eisenstadt, S. N. with Curelaru, M. (1976) *The Form of Sociology – Paradigms and Crises* (John Wiley, New York)

Eliot, T. S. (1960) 'The Influence of Landscape Upon the Poet', *Daedalus*, 89, 421-2

Eliot-Hurst, M. (1980) 'Geography, Social Science and Society, Towards a De-definition', *Australian Geographical Studies*, 18

Emmanuel, A. (1972) *Unequal Exchange: A Study of the Imperialism of Trade* (Monthly Review Press, New York)

Engels, F. (1941) *Ludwig Feuerbach and the Outcome of Classical German Philosophy* (International Publishers, New York)

Entrikin, J. N. (1977) 'Contemporary Humanism in Geography', *Ann. Ass. Am. Geogr.*, 66, 615-32

Febvre, L. A. (1932) *Geographical Introduction to History* (Kegan Paul, Trench and Trubner, London)

Fell, B. (1976) *America B.C.* (Quadrangle/New York Times Book Co., New York)

Fenneman, N. M. (1919) 'The Circumference of Geography', *Ann. Ass. Amer. Geogr.*, 9, 3-11

Feyerabend, P. (1970) 'Against Method: Outline of an Anarchistic Theory of Knowledge', in M. Radner and S. Winokur (eds.) *Minnesota Studies in the Philosophy of Science*, vol. 4 (University of Minnesota Press, Minneapolis)

Fisher, E. and Marke, F. (1971) *The Essential Marx* (Monthly Review Press, New York)

Folk, G. E., Jr. (1974) *Textbook of Environmental Physiology*, 2nd ed. (Lea and Febiger, Philadelphia)

Foote, D. C. and Greer-Wootten, B. (1968) 'An Approach to Systems Analysis in Cultural Geography', *Prof. Geogr.*, 20, 86-91

Frank, A. G. (1969) *Latin America: Underdevelopment of Revolution* (Monthly Review Press, New York)

Frazier, J. W. and Budin, M. (1979) 'An Application of Innovative Behavior Theory to the Assessment of a Housing Rehabilitation Program', *Applied Geography Conference*, 2, 88-105

Freeman, T. W. (1967) *The Geographer's Craft* (Manchester University Press, Manchester)

Friedericks, R. W. (1970) *A Sociology of Sociology* (Free Press, New York)

Galois, R. (1979) *Social Structure in Space: The Making of Vancouver, 1886-1901* (Ph.D. Dissertation, Simon Fraser University, Burnaby)

Gardiner, P. (ed.) (1959) *Theories of History* (Free Press, New York)

George, N. (1966) *Existentialism Versus Marxism* (Dell Publishing Co., New York)

Giacometti, A. (1964) *Alberto Giacometti* (Editions Galerie Beyeler, Basel)

Gillespie, C. (1960) *The Edge of Objectivity* (Princeton University Press, Princeton, NJ)

Glacken, C. (1967) *Traces on the Rhodian Shore: Nature and Culture in Western Thought from Ancient Times to the End of the 18th Century* (University of California Press, Berkeley)

Godolier, M. (1977) *Perspectives in Marxist Anthropology* (Cambridge University Press, Cambridge)

Goethe, J. W. (1970) *Italian Journey 1786-1788* (Penguin, Harmondsworth)

Goldman, R. R. (1973) 'Environmental Limits, Their Prescription and Proscription', *International Journal of Environmental Studies*, 5, 193-204

Goldschmidt, W. R. (1966) *Comparative Functionalism: An Essay in*

Anthropological Theory (University of California Press, Berkeley)

Goldstein, L. J. (1970) 'Collingwood's Theory of Historical Knowing', *History and Theory* 9, 3-36

Golledge, R. G. (1979) 'The Development of *Geographical Analysis*', *Ann. Ass. Am. Geogr.*, 69, 151-4

Golledge, R. and Amadeo, D. (1968) 'On Laws in Geography', *Ann. Ass. Am. Geogr.*, 58, 160-74

Golledge, R. J., Brown, L. A. and Williamson, F. (1972) 'Behavioural Approaches in Geography: An Overview', *The Australian Geographer*, 12, 59-79

Goode, W.J. (1973) *Explorations in Social Theory* (Oxford University Press, New York)

Gottman, J. (1951-2) 'Geography and International Relations', *World Politics*, 3, 153-73

Gould, P. (1979) 'Geography 1957-1977: The Augean Period', *Ann. Ass. Am. Geogr.*, 69, 139-50

Gould, P. and White, R. (1974) *Mental Maps* (Penguin Books, Baltimore)

Gouldner, A. W. (1959) 'Reciprocity and Autonomy in Functional Theory' in L. Gross (ed.) *Symposium on Sociological Theory* (Row, Peterson, Illinois)

—— (1970) *The Coming Crises of Western Sociology* (Basic Books, New York)

Gowans, A. (1974) *On Parallels in Universal History* (Institute for the Study of Universal History, Watkins Glen)

Greene, J. C. (1959) *The Death of Adam: Evolution and its Impact on Western Thought* (Iowa State University Press, Ames)

Gregory, D. (1978) *Ideology, Science and Human Geography* (St. Martins Press, New York)

Grossman, L. (1977) 'Man-Environment Relationships in Anthropology and Geography', *Ann. Ass. Am. Geogr.*, 67, 126-45

Guelke, L. (1971) 'Problems of Scientific Explanation in Geography', *The Canadian Geographer*, 15, 38-53

——(1974) 'An Idealist Alternative in Human Geography', *Ann. Ass. Am. Geogr.*, 64, 193-202

—— (1977a) 'The Role of Laws in Human Geography', in *Progress in Human Geography* (Edward Arnold, London), 376-86

—— (1977b) 'Regional Geography', *Prof. Geogr.*, 29, 1-7

Haas, E. B. (1964) *Beyond the Nation State: Functionalism and International Organization* (Stanford University Press, Stanford)

Haggett, P. (1966) *Locational Analysis in Human Geography* (St. Martins Press, New York)

Haggett, P., Cliff, A.D. and Frey, A. (1977) *Locational Analysis in Human Geography*, vols. I and II, 2nd ed. (Harper and Row, New York)

Hall, E. T. (1969) *The Hidden Dimension* (Doubleday, Garden City, NY)

Hardin, G. (ed.) (1969) *Population, Evolution, and Birth Control*, 2nd ed. (W. H. Freeman, San Francisco)

Harré, R. (1970) *The Principals of Scientific Thinking* (University of Chicago, Chicago)

Harris, E. E. (1969) *Fundamentals of Philosophy: A Study of Classical Texts* (Holt, Rinehart and Winston, New York)

Hartshorne, R. (1950) 'The Functional Approach in Political Geography', *Ann. Ass. Am. Geogr.*, 40, 95-130

—— (1961) *The Nature of Geography* (Association of American Geographers, Lancester, PA)

Harvey, D. (1969) *Explanation in Geography* (St. Martins Press, London)

—— (1972) 'Revolutionary and Counter Revolutionary Theory in Geography and the Problem of Ghetto Formation', *Antipode*, 4, 2, 1-12

—— (1973) *Social Justice and the City* (Johns Hopkins, Baltimore)

—— (1974) 'Population, Resources, and the Idealogy of Science', *Economic Geography*, 50, 256-77

—— (1975) 'The Geography of Capitalist Accumulation: A Reconstruction of the Marxian Theory' *Antipode*, 7, 9-21

—— (1978) 'The Urban Process under Capitalism: A framework for Analysis', *International Journal of Urban and Regional Research*, 2, 101-31

Heidegger, M. (1971) *Poetry, Language, Thought* (Harper and Row, New York)

Heinemann, F. H. (1958) *Existentialism and the Modern Predicament* (Harper and Row, New York)

Hempel, C. G. (1959) 'The Logic of Functional Analysis' in L. Gross (ed.) *Symposium on Sociological Theory* (Row, Peterson, Evanston, Ill.) 271-307

Hesse, M. (1966) *Models and Analogues in Science* (University of Notre Dame, Notre Dame, Ill.)

Hicks, D. (1938) *Critical Realism* (Macmillan, London)

Hindess, B. and Hirst, P. (1975) *Pre-Capitalist Modes of Production* (Routledge and Kegan Paul, London)

—— (1977) *Mode of Production and Social Formation* (Macmillan, London)

Hofstadter, R. (1955) *Social Darwinism in American Thought*, rev. ed. (Beacon Press, Boston)

Homans, G. C. (1964) 'Bringing Men Back In', *American Sociological Review*, 29, 809-18

Hsu, Shin-Yi, (1969) 'The Cultural Ecology of the Locust Cult in Traditional China', *Ann. Ass. Am. Geogr.*, 59, 731-52

Hudson, J. (1969) 'A Location Theory for Rural Settlement', *Ann. Ass. Am. Geogr.*, 52, 365-81

Huntington, E. (1919) *The Red Man's Continent: A Chronicle of Aboriginal America* (Yale University Press, New Haven, Conn.)

—— (1921) 'Air Control as a Means of Reducing the Post-Operative Death Rate', *American Journal of Surgery: Anaesthesia Supplement*, 35, 82-90 and 98-100

—— (1923) *Earth and Sun* (Yale University Press, New Haven, Conn.)

—— (1924) *Civilization and Climate*, 3rd ed. (Yale University Press, New Haven, Conn.)

—— (1945) *Mainsprings of Civilization* (Wiley, New York)

Husserl, E. (1970) *The Crisis of the European Sciences and Transcendental Phenomenology* (Northwestern UP, Evanston)

Ihde, D. (1977) *Experimental Phenomenology* (Putnam, New York)

Isajiw, W. W. (1968) *Causation and Functionalism in Sociology* (Shocken Books, New York)

Itoh, S. (1974) *Physiology of Cold-Adapted Man* (Hokkaido University School of Medicine, Sapporo, Japan)

James, P. (1972) *All Possible Worlds: A History of Geographical Ideas* (Bobbs-Merrill, Indianapolis)

Bibliography

215

James, P. E. and Jones, C. F. (eds.) (1954) *American Geography: Inventory and Prospect* (Syracuse University Press, Syracuse, NY)

James P. E. and Martin, G. J. (1979) *The AAG, The First Seventy-Five Years: 1904-1979* (Association of American Geographers, Washington)

James, W. (1904) 'Review of F. C. S. Schiller's Humanism', *Nation*, 78, 175-6

—— (1932) *A Pluralistic Universe: Hibbert Lectures on the Present Situation in Philosophy* (Longmans, Green, New York)

Jarrett, J. L. and McMurrin, S. M. (1954) 'Pragmatism' in Jarrett and McMurrin (eds.) *Contemporary Philosophy*, (Henry Holt, New York), 253-7

Jarvie, I. C. (1973) *Functionalism* (Burgess Publishing Co., Minneapolis, Minn.)

Jaspers, K. (1957) *Man in the Modern Age* (Doubleday, Garden City, New York)

—— (1969) *Philosophy* (University of Chicago Press, Chicago), vols. 1 and 2

Jay, M. (1973) *The Dialectical Imagination: A History of the Frankfurt School and the Institute of Social Research, 1923-1950* (Little, Brown, Boston)

Jessop, B. (1972) *Social Order, Reform, and Revolution* (Macmillan, London)

Joerg, W. L. G. (1922) 'Recent Geographical Works in Europe' *Geogr. Rev.*, 12, 431-84

Joergensen, J. (1951) *The Development of Logical Empiricism* (University of Chicago Press, Chicago)

Johnson, R. J. (1977) 'National Sovereignty and National Power in European Institutions', *Environment and Planning, A*, 9, 569-78

Kasperson, R. E. and Minghi, J. V. (eds.) (1969) *The Structure of Political Geography* (Aldine, Chicago)

Keat, R. and Urry, J. (1975) *Social Theory as Science* (Routledge and Kegan Paul, London)

Kennedy, W. (1973) 'The Yellow Trolley Car in Barcelona and Other Visions: A Profile of Gabriel Garcia Marques', *Atlantic Monthly*, January, 50-9

King, L. J. (1976) 'Alternatives to a Positive Economic Geography', *Ann. Ass. Am. Geogr.*, 66, 293-308

Kirk, W. (1951) 'Historical Geography and the Concept of the Behavioral Environment', *Indian Geographical Journal*, Silver Jubilee Volume, 152-60

Kolb, W. L. (1961) 'Images of Man and the Sociology of Religion', *Journal for the Scientific Study of Religion*, 1, 5-22

Kollmorgen, W. (1954) 'And Deliver Us From Big Dams', *Land Economics*, 30, 333-46

Kriesel, K. M. (1968) 'Montesquieu: Possibilistic Geographer', *Ann. Ass. Am. Geogr.*, 58, 557-74

Kropotkin, P. (1972) *Mutual Aid* (New York University Press, New York)

Krueger, A. P. and Reed, E. J. (1976). 'Biological Impact of Small Air Ions', *Science*, 193, 1209-13

Kuhn, T. S. (1961) 'The Function of Measurement in Modern Physical Science', *Isis*, 52, 161-93

—— (1970a) *The Structure of Scientific Revolutions* (University of Chicago, Chicago)

—— (1970b) 'Logic of Discovery or Psychology of Research?', in Imre Lakatos (ed.), *Criticisms and the Growth of Knowledge* (University Press, Cambridge), 1-23

—— (1970c) 'Reflections on my Critics', in Imre Lakatos (ed.), *Criticisms and the Growth of Knowledge* (University Press, Cambridge), 231-78

Lamb, H. B. (1972) *Vietnam's Will to Live* (Monthly Review Press, New York and London)

Lamb, H. H. (1972-7) *Climate: Present, Past, and Future*, 2 vols. (Methuen, London)

Langton, J. (1974) 'Potentialities and Problems of Adopting a Systems Approach to the Study of Change in Human Geography', in C. Board, R. J. Chorley, P. Haggett and D. R. Stoddart, (eds.) *Progress in Geography* (Edward Arnold, London)

Lappé, F. M. and Collins, J. (1977) *Food First: Beyond the Myth of Scarcity* (Houghton Mifflin, Boston)

Lazlo, E. (1972) *Introduction to Systems Philosophy* (Harper and Row, New York)

Lefebvre, H. (1968) *Dialectical Materialism* (Cape, London)

—— (1972) *La Pensée Marxiste et la Ville* (Casterman, Paris)

Leighly, J. (1976) 'Carl Ortwin Sauer, 1889-1975', *Ann. Ass. Am. Geogr.*, 66, 337-48

Leiss, W. (1974) *The Domination of Nature* (G. Braziller, New York)

LeRoy, L. E. (1971) *Times of Feast, Times of Famine: A History of Climate Since the Year 1000* (Doubleday, Garden City, NY); translated by Barbara Bray, from French edition, 1967

Lessing, D. (1973) *The Golden Notebook* (Panther Books, St. Albans)

Levi-Strauss, C. (1963) *Structural Anthropology* (Basic Books, New York)

Ley, D. (1977) 'Social Geography and the Taken-for-Granted World', *Trans. Inst. Br. Geogr.*, new series, 2, 498-512

Ley, D. and Samuels, M. S. (eds.) (1978) *Humanistic Geography: Prospects and Problems* (Maaroufa Press, Chicago)

Mabogunje, A. L. (1970) 'Systems Approach to a Theory of Rural-Urban Migration', *Geographical Analysis*, 2, 1-8

McGregor, M. (1972) 'Evaluation of Huntington's Ozone Hypothesis as a Basis for His Cyclonic Man Theory' (unpublished MA thesis, University of Kansas, Lawrence)

McNeill, H. (1974) *The Shape of European History* (Oxford University Press, London)

Malinowski, B. (1920) 'Anthropology', *Encyclopedia Brittanica*, First Supplementary Volume (Encyclopedia Brittanica, New York)

—— (1926) *Argonauts of the Western Pacific* (Routledge, London)

Malinowski, B. (1936) 'Anthropology', *Encyclopedia Britannica* (Encyclopedia Britannica, New York)

—— (1944) *A Scientific Theory of Culture and Other Essays* (University of North Carolina Press, Chapel Hill)

—— (1954) *Magic, Science and Religion* (Doubleday-Anchor, New York)

Mandelsohn, H. (1974) 'Behaviourism, Functionalism, and Mass Communications Policy', *Public Opinion Quarterly*, 38, 379-89

Marx, K. (1967) *Capital*, 3 vols. (International Publishers, New York)

—— (1970) *A Contribution to the Critique of Political Economy* (International Publishers, New York)

Marx, K. and Engels, F. (1975) *The German Ideology in Marx and Engels, Collected Works*, vol. 5 (International Publishers, New York)

Masterman, M. (1970) 'The Nature of a Paradigm', in Imre Lakatos (ed.),

Criticism and the Growth of Knowledge (University Press, Cambridge), 59-89

Mayfield, R. C. and English, P. W. (eds.) (1972) *Man, Space, and Environment* (Oxford University Press, New York)

Mercer, D. C. and Powell, J. M. (1972) *Phenomenology and Related Non-Positivistic Viewpoints in the Social Sciences* (Department of Geography, Monash University, Melbourne, Australia)

Merleau-Ponty, M. (1962) *Phenomenology of Perception* (Humanities Press, New York)

Merton, R. K. (1948) 'Discussion of Parsons "The Position of Sociological Theory"', *American Sociological Review*, 13, 164-8

—— (1957) *Social Theory and Social Structure*, (Free Press, Glencoe, Ill.)

—— (1976) *Sociological Ambivalence* (Free Press, New York)

Mesarovic, M. and Pestel, E. (1974) *Mankind at the Turning Point* (New American Library, New York)

Mills, C. W. (1951) *White Collar: The American Middle Class* (Oxford University Press, New York)

Mitrany, D. (1977) *The Functional Theory of Politics* (St. Martins Press, London)

Morrill, R. J. (1969) 'Geography and the Transformation of Society', Part 1, *Antipode*, 1, 6-9

—— (1970) 'Geography and the Transformation of Society, Part 2, *Antipode*, 2, 4-10

—— (1973) 'Ideal and Reality in Reapportionment', *Ann. Ass. Am. Geogr.*, 63, 463-77

Murphey, R. (1967) 'Man and Nature in China', *Modern Asian Studies*, 1, 313-33

Nagel, E. (1961) *The Structure of Science* (Routledge and Kegan Paul, London)

Natanson, M. (ed.) (1963) *Philosophy of the Social Sciences: A Reader* (Random House, New York)

Neurath, O. (ed.) (1938-69) *International Encyclopedia of Unified Science*, vols. I and II (University of Chicago Press, Chicago)

Nevins, R. G., Gonzalez, R. R., Nishi, Y. and Gagge, A. P. (1975) 'Effect of Changes in Ambient Temperature and Level of Humidity on Comfort and Thermal Sensations', *Transactions of American Society of Heating, Refrigerating, and Air-Conditioning Engineers*, no. 2370 RP-144

Palloix, C. (1977) 'The Self-Expansion of Capital on a World Scale', *The Review of Radical Political Economics*, 9, 3-28

Panofsky, E. (1972) *Studies in Iconology* (Oxford University Press, New York)

Parker, G. (1974) 'The Logic of Unity: A Geography of the European Economic Community (Longman, London)

Parsons, H. L. (ed.) (1977) *Marx and Engels on Ecology* (Greenwood Press, Westport, Connecticut)

Parsons, T. (1949) *Essays in Sociological Theory Pure and Applied* (Free Press, Glencoe, Ill.)

—— (1977) *Social Systems and the Evolution of Action Theory* (Free Press, New York)

Passmore, J. (1966) *A Hundred Years of Philosophy* (Basic Books, New York)

Peet, R. (1977) 'The Development of Radical Geography in the United States', in R. Peet (ed.), *Radical Geography* (Maaroufa Press, Chicago), 6-30

—— (1978a) 'The Geography of Human Liberation', *Antipode*, 10, 119-34

—— (1978b) 'The Dialectics of Radical Geography: A Reply to Gordon Clark and Michael Dear', *The Prof. Geogr.*, 30, 360-4

—— (1979a) 'Societal Contradiction and Marxist Geography', *Ann. Ass. Am. Geogr.*, 69, 164-9

—— (1979b) *Radical Geography* (Maaroufa Press, Chicago)

Perpillou, A. V. (1966) *Human Geography* (Wiley, New York)

Popper, K. (1961) *The Logic of Scientific Discovery* (Basic Books, New York)

Post, D. (1977) *The Last Great Subsistence Crisis in the Western World* (John Hopkins University Press, Baltimore)

Pred, A. R. (1974) *Job-Providing Organizations and Systems of Cities*, Resource Paper No. 27 (Association of American Geographers, Washington)

Radcliffe-Brown, A. R. (1952) *Structure and Function in Primitive Society* (Cohen and West, London)

—— (1977) 'Function, "Meaning" and "Functional Consistency" ', in Adam Kuper (ed.), *The Social Anthropology of Radcliffe-Brown* (Routledge and Kegan Paul, New York), 43-52

Radnitzky, G. (1970) *Contemporary Schools of Metascience*, vols I and II (2nd ed., Goteborg)

Ravetz, J. R. (1973) *Scientific Knowledge and its Social Problems* (Oxford University Press, New York)

Relph, E. (1970) 'An Inquiry into the Relations Between Phenomenology and Geography', *Canadian Geographer*, 15, 181-92

—— (1976a) 'The Phenomenological Foundations of Geography', *Discussion Paper No. 21*, Dept. of Geography (University of Toronto, Toronto)

—— (1976b) *Place and Placelessness* (Pion, London)

Rescher, N. (1978) *Pierce's Philosophy of Science: Critical Studies in his Theory of Induction and Scientific Method* (Notre Dame Press, Notre Dame)

Rickert, H. (1962) *Science and History: A Critique of Positivist Epistemology*, A. Goddard (ed.) (Van Nostrand, Princeton NJ)

Rickman, H. P. (1961) *Meaning in History: W. Dilthey's Thoughts on History and Society* (Allen and Unwin, London)

Ritzer, G. (1975) *Sociology: A Multiple Paradigm Science* (Allyn and Bacon, Boston)

Roberts, W. O. and Lansford, H. (1979) *The Climate Mandate* (W. H. Freeman, San Francisco)

Rohles, F. H., Jr. (1975) 'Humidity, Human Factors, and the Energy Shortage', *American Society of Heating, Refrigerating, and Air-Conditioning Engineers Journal*, 17, 38-40

Rostlund, E. (1956) 'Twentieth-Century Magic', *Landscape*, 5, 23-6 (reprinted in Philip Wagner and Marvin Mikesell, *Readings in Cultural Geography* (University of Chicago Press, Chicago, 1962), 48-53

Rubinoff, L. (1970) *Collingwood and the Reform of Metaphysics* (Toronto University Press, Toronto)

Rucker, D. (1969) *The Chicago Pragmatist* (University of Minnesota Press, Minneapolis)

Ruskin, J. (1905) *Modern Painters*, vol III (George Routledge and Sons, London)

Russell, B. (1959) *Wisdom of the West* (Doubleday, Garden City)

Samuels, M. S. (1978) 'Existentialism and Human Geography', in D. Ley and M. S. Samuels (eds.) *Humanistic Geography* (Maaroufa Press, Chicago), 22-40

—— (1979) 'The Biography of the Landscape', in D. Meinig (ed.), *The Interpretation of Ordinary Landscapes* (Oxford University Press, London)

Santos, M. (1977) 'Society and Space: Social Formation as Theory and Method', *Antipode*, 9, 3-13

Sartre, J. P. (1946) *L'existentialisme est un Humanisme* (Editions Nagel, Paris)
—— (1966) *Being and Nothingness* (Washington Square Press, New York)

Sauer, C. O. (1925) 'The Morphology of Landscape', *University of California Publications in Geography*, 2, 19-53
—— (1941) 'Foreword to the Historical Geography', *Ann. Ass. Amer. Geogr.*, 31, 1-24
—— (1956) 'The Education of a Geographer', *Ann. Ass. Am. Geogr.*, 45, 287-99
—— (1971) 'The Formative Years of Ratzel in the United States', *Ann. Ass. Am. Geogr.*, 61, 245-54
—— (1974) 'The Fourth Dimension of Geography', *Ann. Ass. Amer. Geogr.*, 64, 2, 189-92

Schaefer, F. K. (1953) 'Exceptionalism in Geography', *Ann. Ass. Amer. Geogr.*, 43, 226-49

Scheffler, I. (1974) *Four Pragmatists* (Humanities Press, New York)

Scheler, M. (1970) *Man's Place in Nature* (Noonday Press, New York)

Schmidt, A. (1971) *The Concept of Nature in Marx* (New Left Books, London)

Schneider, S. with Mesirow, L. (1976) *The Genesis Strategy* (Plenum Press, New York)

Schumacher, E. F. (1977) *A Guide for the Perplexed* (Fitzhenry and Whiteside, Toronto)

Schutz, A. (1962) *Collected Papers*, vols. I and II (Martinus Nijhoff, The Hague)

Seamon, D. (1979) *A Geography of the Lifeworld* (St. Martin's Press, New York)

Semple, E. C. (1911) *Influences of Geographic Environment on the Basis of Ratzel's System of Anthropo-Geography* (Henry Holt, New York)

Sewell, J. P. (1966) *Functionalism and World Politics* (Princeton University Press, Princeton, NJ)

Slonim, N. (ed.) (1974) *Environmental Physiology* (C. V. Mosby, St. Louis)

Smith, N. (1967) *Medieval Art* (D. W. Braun, Iowa)

Soja, E. W. (1968) 'Communications and Territorial Integration in East Africa: An Introduction to Transaction Flow Analysis', *The East Lakes Geographer*, 4, 39-57
—— (1978) '"Topian" Marxism and Spatial Praxis: A Reconsideration of the Political Economy of Space', paper presented to the Annual Meeting of the Association of American Geographers, New Orleans

Sommer, R. (1969) *Personal Space: The Behavioral Basis of Design* (Prentice Hall, Englewood Cliffs, NJ)

de Souza, A. R. and Porter, P. W. (1974) 'The Underdevelopment and Modernization of the Third World', *Research Paper 28* (Association of American Geographers, Washington)

Stewart, D. and Mickunas, A. (1974) *Exploring Phenomenology: A Guide to the Field of Literature* (American Library Association, Chicago)

Stohr, W. B. (1974) *Interurban Systems and Regional Economic Development*, Commission on College Geography, Resource Paper No. 26 (Association of American Geographers, Washington)

Suppe, F. (ed.) (1977) *The Structure of Scientific Theories*, 2nd ed. (University of Illinois Press, Urbana)

Szentes, T. (1971) *The Political Economy of Underdevelopment* (Akadémiai Kiadó, Budapest)

Taaffe, E. (1974) 'The Spatial View in Context', *Ann. Ass. Am. Geogr.*, 64,
    1-16
Taylor, G. (ed.) (1951) *Geography in the Twentieth Century* (Methuen, London)
Taylor, P. and Groom. A. J. R. (1975) 'Functionalism and International
    Relations', in P. Taylor and A. J. R. Groom (eds.) *Functionalism: Theory
    and Practice in International Relations* (Crane, Russak, New York), 1-6
Thayer, H. S. (1968) *Meaning and Action. A Critical History of Pragmatism*
    (Bobbs-Merrill, New York)
Titus, H. H. (1964) *Living Issues in Philosophy*, 4th ed. (American Book,
    New York)
Tolman, E. C. (1932) *Purposive Behavior in Animals and Men* (D. Appleton-
    Century, New York)
Tromp, S. (1967) 'Biometeorology' (114-22) and 'Climate and Evolution'
    (161-6), in Rhodes, W. Fairbridge (ed.) *The Encyclopedia of Atmospheric
    Sciences and Astrogeology* (Reinholdt Publishing Corp., New York)
Tse-Tung, Mao (1965) *Selected Works* (Foreign Language Press, Peking)
Tuan, Yi-Fu (1971) 'Geography, Phenomenology and the Study of Human
    Nature', *Canadian Geographer*, 15, 181-92
——— (1974) *Topophilia: A Study of Environmental Perception, Attitudes
    and Values* (Prentice-Hall, Englewood Cliffs, NJ)
——— (1977) *Space and Place* (University of Minnesota Press, Minneapolis)
Turner, J. H. (1974) *The Structure of Sociological Theory* (Dorsey Press,
    Homewood, Illinois)
Tymieniecka, Anna-Teresa (1962) *Phenomenology and Science in Contemporary
    European Thought* (Noonday Press, New York)
Ullman, E. J. (1953) 'Human Geography and Area Research', *Ann. Ass. Amer.
    Geogr.*, 43, 54-66
Van den Berghe, Pierre L. (1967) 'Dialectic and Functionalism: Toward a
    Synthesis', in N. J. Demerath III and Richard A. Peterson (eds.), *System,
    Change, and Conflict* (Free Press, New York), 293-306
Van Valkenburg, S. and Held, C. C. (1952) *Europe*, 2nd ed. (John Wiley,
    New York)
Wail, E. (1965) 'Science in Modern Culture, or the Meaning of Meaninglessness',
    Daedalus, 94, 171-89
Wallerstein, I. (1974) 'The Rise and Future Demise of the World Capitalist
    System', *Comparative Studies in Society and History*, 16, 387-415
Walmsley, D. J. (1974) 'Positivism and Phenomenology in Human Geography',
    Canadian Geographer, 18, 95-107
Walsh, D. (1973) 'Functionalism and Systems Theory', in Paul Filmer *et al.* (eds.),
    New Directions in Sociological Theory (MIT Press, Cambridge, Mass),
    57-76
Wanklyn, H. (1961) *Friedrich Ratzel: A Biographical Memoir and Bibliography*
    (Cambridge University Press, Cambridge)
Wheatley, P. (1976) 'The Suspended Pelt, Reflections of a Discarded Model of
    Spatial Structure', mimeograph paper (to be published by the University of
    Michigan)
Whitehead, A. (1929) *Process and Reality* (Cambridge University Press,
    Cambridge)
Whitehead, A. N. and Russell, B. (1910-13) *Principia Mathematica*, vols. I-III
    (Cambridge University Press, Cambridge)

Wiener, P. P. (ed.) (1966) *C. S. Peirce: Selected Writings* (Dover, New York)

Wilbanks, T. and Symanski, R. (1968) 'What is Systems Analysis?', *Prof. Geogr.*, 20, 81-91

Wild, J. (1963) *Existence and the World of Freedom* (Prentice Hall, Englewood Cliffs)

Willer, D. and Willer, J. (1973) *Systematic Empiricism: Critique of a Pseudoscience* (Prentice Hall, Englewood Cliffs, NJ)

Willingform, D. (1936) 'A New Yorker's Idea of the United States of America', *Saturday Review of Literature*, 15

Wittfogel, K. (1957) *Oriental Despotism* (Yale University Press, New Haven)

Wittgenstein, L. (1963) *Tractatus Logico-Philosophicus* (German/English edition, Routledge and Kegan Paul, London)

Woodger, J. H. (1939) *The Technique of Theory Construction* (University of Chicago Press, Chicago)

Worsley, P. (1978) 'Whither Geomorphology?' *Area*, 11, 97-101

Wright, J. (1966) *Human Nature in Geography* (Harvard University Press, Cambridge)

Wyon, D. P. (1974) 'The Effects of Moderate Heat Stress on Typewriting Performance', *Ergonomics*, 17, 309-18

Zelinsky, W. (1973) 'Beyond the Exponentials: The Role of Geography in the Great Transition', *Economic Geography*, 46, 498-535

—— (1975) 'The Demigod's Dilemma', *Ann. Ass. Am. Geogr.*, 65, 123-43

Zimmermann, R. E. (1964) *Introduction to World Resources*, H. L. Hunker (ed.) (Harper and Row, New York)

Zinn, H. (1967) *Vietnam: The Logic of Withdrawal* (Beacon Press, Boston)

NOTES ON CONTRIBUTORS

John E. Chappell, Jr, San Luis Obispo, California.

John W. Frazier, Department of Geography, SUNY Binghamton, Binghamton, New York.

Edward M. W. Gibson, Department of Geography, Simon Fraser University, Burnaby, British Columbia.

Leonard Guelke, Department of Geography, University of Waterloo, Waterloo, Ontario.

Milton E. Harvey, Department of Geography, Kent State University, Kent, Ohio.

Michael R. Hill, Department of Earth Sciences, Iowa State University, Ames, Iowa.

Brian P. Holly, Department of Geography, Kent State University, Kent, Ohio.

James V. Lyons, Department of Geography, Valparaiso University, Valparaiso, Indiana.

Richard J. Peet, Graduate School of Geography, Clark University, Worcester, Massachusetts.

Edward C. Relph, Department of Geography, University of Toronto, Toronto, Ontario.

Marwyn S. Samuels, Department of Geography, University of British Columbia, Vancouver, British Columbia.

INDEX

Printed and bound by CPI Group (UK) Ltd, Croydon, CR0 4YY

01/11/2024

01782633-0001